LIVING UNDERWATER

LIVING UNDERWATER

Farooq Hussain

Studio Vista

Produced by November Books Limited, 23–9 Emerald Street, London, WC1N 3QL

Published by Studio Vista Limited, Blue Star House, Highgate Hill, London, N19

Text set by Yendall and Company Limited, Riscatype House, 22–5 Red Lion Court, Fleet Street, London, EC4

Printed by Taylor Garnett Evans and Company Limited, Greycaine Road, Bushey Mill Lane, Watford, Herts.

Bound by Dorstel Press Limited, West Road, Templefields, Harlow, Essex

SBN 289.70060.4 (hardbound)
289.70059.0 (paperback)

Designed by Tom Carter.
House Editor: Celia Phillips.
Copy preparation: John Leath.

Contents

To my parents and brother

Acknowledgements.

I am greatly indebted to Keith and Gail Critchlow for help, encouragement, consolation and affection over the last four years; Admiral Stanley Miles, RN, for his personal interest in my work and the trouble taken in reading many of my articles; Commdt. Cousteau, for his generous help with my work, and particularly to Jean-Michel Cousteau in appreciation of work done together; Paul Oliver for his constructive interest in my work; Celia Phillips and Tom Carter for their help in preparing this book. I should also like to thank the following people who have given their time and valuable criticism to my work, or have generously provided me with material for my research:

Dr Leonard Libber, Head of Physiology, Office of Naval Research, United States Navy, Washington, D.C.
Dr John West, Professor of Medicine and Bioengineering, University of California, San Diego.
Professor Jacques Chouteau, Laboratoire de Physiologie Appliquée, Faculté des Sciences de Marseille.
H. V. Hempleman, Superintendent, Royal Naval Physiological Laboratory, Alverstoke, Hants.
Dr Joannes Kylstra, Professor of Physiology, Associate Professor of Medicine, Duke University, North Carolina.
Dr G. D. Campbell, physician, Durban, South Africa.
Commdt. Jean Alinat, Director of the Centre d'Etudes Marine Avancées, Marseilles.
Commander George Bond, USN, Project Sealab, Long Beach, California.
Commander Scott Carpenter, USN Project Sealab.
Dr Jacques Piccard, Lausanne, Switzerland.
Raymond Vaissière, Musée Oceanographique de Monaco.
Bill Weston.
Alex Double.

‘Menfish’

by

Commdt. J.-Y. Cousteau

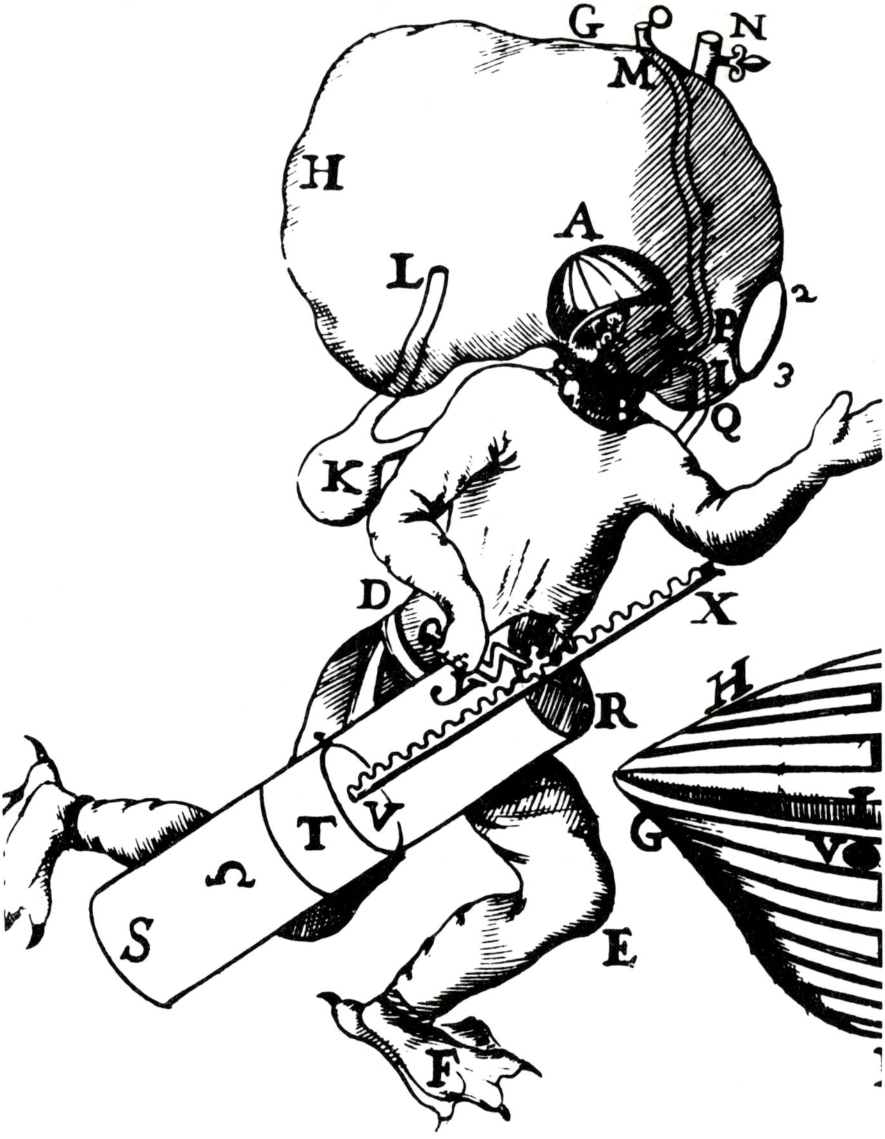

An early attempt at the aqua-lung concept.

One morning in June 1943, I went to the railway station at Bandol on the French Riviera to collect a wooden case expressed from Paris. In it was a new and promising device, the result of years of struggle and dreams: an automatic compressed-air diving lung conceived by Emile Gagnan and myself. I rushed it to Villa Barry where my diving comrades, Philippe Tailliez and Frédéric Dumas, were waiting. No children ever opened a Christmas present with more excitement than we did when we unpacked the first 'aqualung'. If it worked, diving could be revolutionised.

We found an assembly of three moderate sized cylinders of compressed air, linked to an air regulator the size of an alarm clock. The bottles contained air condensed to one hundred and fifty times atmospheric pressure. From the regulator there extended two tubes, joining on to a mouthpiece. With this equipment harnessed to the back, a watertight glass mask over the eyes and nose, and rubber foot fins, we intended to make unencumbered flights in the depths of the sea.

We hurried to a sheltered cove which would conceal our activity from curious bathers and Italian occupation troops. I checked the air pressure. It was difficult to contain my excitement and discuss calmly the plan of the first dive. Dumas, the best goggle diver in France, would stay on shore keeping warm and rested, if necessary, to dive to my aid. My wife, Simone, would swim out on the surface with a schnorkel breathing-tube and watch me through her submerged mask. If she signalled anything had gone wrong, Dumas could dive to me in seconds. 'Didi', as he was known on the Riviera, could 'skin dive' to sixty feet.

My friends harnessed the three-cylinder block on my back with the regulator riding at the nape of my neck and the hoses looped over my head. I spat on the inside of my shatterproof glass mask and rinsed it in the surf, so that mist would not form inside. I moulded the soft rubber flanges of the mask tightly over forehead and cheekbones and then I fitted the mouthpiece under my lips and gripped the nodules between my teeth. A vent the size of a paper clip was to pass my inhalations and exhalations beneath the sea. Staggering under the fifty-pound apparatus I walked with a Charlie Chaplin waddle into the sea.

The diving lung was designed to be slightly buoyant. I reclined in the chilly water to estimate my compliance with Archimedes' principle that a solid body immersed in liquid is buoyed up by a force equal to the weight of the liquid displaced. Dumas justified me with Archimedes by attaching seven pounds of lead to my belt. I sank gently to the sand. I breathed sweet effortless air. There was a faint whistle when I inhaled and a light rippling sound of bubbles when I breathed out. The regulator was adjusting pressure precisely to my needs.

I looked into the sea with the same sense of trespass that I have felt on every dive. A modest canyon opened below, full of dark green weeds, black sea urchins, and small flowerlike white algæ. Fingerlings browsed in the scene. The sand sloped down into a clear blue infinity. The sun struck so brightly I had to squint. My arms hanging at my sides, I kicked the fins languidly and travelled down, gaining speed, watching the beach reeling past. I stopped kicking and the momentum carried me on a fabulous glide. When I stopped, I slowly emptied my lungs and held my breath. The diminished volume of my body decreased the lifting force of water, and I sank dreamily down. I inhaled a great chestful and retained it. I rose towards the surface.

My human lungs had a new role to play, that of a sensitive ballasting system. I took normal breaths in a slow rhythm, bowed my head, and swam smoothly down to thirty feet. I felt no increasing water pressure, which at that depth is twice that of the surface. The aqualung automatically fed me increased compressed air to meet the new pressure layer. Through the fragile human-lung linings this counter pressure was being transmitted to the blood stream and instantly spread throughout the incompressible body. My brain received no subjective news of the pressure and I was at ease, except for a pain in the middle ear and sinus cavities. I swallowed, as one does when landing in an aeroplane, to open my Eustachian tubes and stopped the pain. (I did not wear ear plugs, a dangerous practice when under water. Ear plugs would have trapped a pocket of air between them and the eardrums. Pressure building up in the Eustachian tubes would have forced my eardrums outward, eventually to the bursting point.)

I reached the bottom in a state of excitement. A school of silvery sars (goat bream), round and flat as saucers, swam in a rocky chaos. I looked up and saw the surface shining like a defective mirror. In the centre of the looking-glass was the trim silhouette of Simone, reduced to a doll. I waved. The doll waved at me.

I became fascinated with my exhalations. The bubbles swelled on the way up through lighter pressure layers, but were peculiarly flattened like mushroom caps by their eager push against the medium. I realised the importance bubbles were to have for us in the dives to come: as long as air bubbled on the surface all was well below. If the bubbles disappeared, there would be anxiety and emergency measures. They roared out of the regulator and kept me company. I felt less alone.

I swam across the rocks and compared myself favourably with the sars. To swim fishlike, horizontally, was the logical method in a medium eight hundred times denser than air. To halt and hang attached to nothing, no lines or air pipe to the surface, was a dream. At night I had often had visions of flying by extending my arms as wings. Now I flew without wings. (Since that first aqualung flight, I have never had a dream of flying.)

I thought of the helmet diver arriving on his ponderous boots at such a depth as this and struggling to walk a few yards, obsessed with his umbilici and his head imprisoned in copper. On 'skin-dives' I had seen him leaning dangerously forward to make a step, clamped in heavier pressure at the ankles than the head, a cripple in an alien land. From this day forward we would swim across miles of country no man had known, free and level, with our flesh feeling what the fish scales know.

I experimented with all possible manoeuvres of the aqualung – loops, somersaults, and barrel rolls. I stood upside down on one finger and burst out laughing – a shrill distorted laugh. Whatever I did, nothing altered the automatic rhythm of air. Delivered from gravity and buoyancy, I flew around in space.

I could attain a speed of almost two knots without using my arms; I soared vertically and passed my own bubbles; I went down to sixty feet. We had been there many times without breathing aids, but we did not know what happened below that boundary. How far could we go with this strange device?

Fifteen minutes had passed since I left the little cove. The regulator lisped in a steady cadence in the ten-fathom layer and I could spend an hour there on my air supply. I determined to stay as long as I could stand the chill. Here were tantalising crevices we had been obliged to pass fleetingly before. I swam inch by inch into a dark narrow tunnel, scraping my chest on the floor and banging the air tanks on the ceiling. In such situations a man is of two minds. One urges him on towards mystery and the other reminds him that, if he will

The Cousteau family on an underwater excursion near Sanary-sur-Mer.

use it, he is a creature with good sense that can keep him alive. I bounced against the ceiling. I had used one-third of my air and was getting lighter. My brain complained that this foolishness might sever my air hoses. I turned over and hung on my back.

The roof of the cave was thronged with lobsters. They stood there like great flies on a ceiling. Their heads and antennæ were pointed towards the cave entrance. I breathed lesser lungsful to keep my chest from touching them. Above water was occupied, ill-fed France. I thought of the hundreds of calories a diver loses in cold water. I selected a pair of one-pound lobsters and carefully plucked them from the roof without touching their stinging spines. I carried them towards the surface.

Simone had been floating, watching my bubbles wherever I went. She swam down towards me. I handed her the lobsters and went down again as she returned to the surface. She came up under a rock which bore a torpid Provençal citizen with a fishing rod. He saw a blonde girl emerge from the combers with lobsters wriggling in her hands. She said, 'Could you please watch these for me?' and put them on the rock. The fisherman dropped his rod.

Simone made five more surface dives to take lobsters from me and carry them to the rock. I surfaced in the cove, out of the fisherman's sight. Simone claimed her lobster swarm. She said, 'Keep one for yourself, monsieur. They are very easy to catch if you do as I did.'

Lunching on the treasures of the dive, Tailliez and Dumas questioned me on every detail. We revelled in plans for the aqualung. Tailliez pencilled the tablecloth and announced that each yard of depth we claimed in the sea would open to mankind 300,000 cubic kilometers of living space. Tailliez, Dumas, and I had come a long way together. We had been eight years in the sea as goggle divers. Our new key to the hidden world promised wonders. We recalled the beginning. . . .

Our first tool was the underwater goggle, a device that was known centuries ago in Polynesia and Japan, was used by 16th-century Mediterranean coral divers, and has been rediscovered about every decade in the last fifty years. The naked human eye, which is almost blind under water, can see clearly through watertight spectacles.

One Sunday morning in 1936 at Le Mourillon, near Toulon, I waded into the Mediterranean and looked into it through Fernez goggles. I was a regular Navy gunner, a good swimmer interested only in perfecting my crawl style. The sea was merely a salty obstacle that burned my eyes. I was astounded by what I saw in the shallow shingle at Le Mourillon – rocks covered with green, brown, and silver forests of algæ and fishes unknown to me, swimming in crystal-clear water. Standing up to breathe I saw a trolley-bus, people, electric street-lights. I put my eyes under again and civilisation vanished with one last bow. I was in a jungle never seen by those who floated on the opaque roof.

Sometimes we are lucky enough to know that our lives have been changed, to discard the old, embrace the new, and run headlong down an immutable course. It happened to me at Le Mourillon on that summer's day, when my eyes were opened on the sea.

J.-Y. COUSTEAU

Preface

Life originated in the primordial oceans of the earth and would probably not have occurred without their presence. The existence of water, and its abundance on the planet, is obvious, but its unique properties and extraordinary behaviour are often taken for granted or go unnoticed. This book is concerned with oceans as the fundamental manifestation of energy on this planet, governing natural phenomena which are accepted as part of everyday experience, and whose integrity and function are crucial to those processes which support and complement life on earth.

The evolution of living organisms from aquatic environments has not changed their dependence on water to support biological processes. Water constitutes more than half of the substance of all living matter, and photosynthesis and respiration, the energy exchange processes of living organisms, rely on the ability of the oceans' surface to absorb the sun's radiation. The condition and behaviour of the oceans are therefore significant factors in the overall ecology of the planet, and any deterioration in their condition will affect life processes.

Modification of the immediate environment enables a living organism to survive for a while, but if the total system is corrupted, then even short term survival is unlikely. Man constantly makes significant changes to the normal ecological function of the planet. When they are implemented without regard to their effect on the ocean system and atmosphere, the inevitable result is a disruption of those processes which are crucial to life.

Biological adaptation to abnormal and alien conditions has been vital to man's evolution; however, the value and extent of future adaptation must obviously have moral implications for those involved with experimenting upon the environment. It is reasonable to accept that exploration is an important part of human behaviour, and that as a result of exploration directives are taken which entail varying degrees of adaptation to particular environmental conditions.

When exploring the ocean, man is faced with considerable problems in terms of physiological adaptation, which sometimes necessitate radical changes in the normal pattern of life. Mechanical and engineering systems have often been suggested as a viable alternative to manned exploration, both for space and undersea operations, though it is obvious to even the most optimistic engineer that the development of equipment with the ability of a man trained in exploration is very far in the future. Millions of years of evolution have been required to achieve the biological sophistication of a human being, and to attempt to duplicate man's functions in part or whole is a waste of resources. Whatever technological systems can be developed to assist man's exploration of the oceans or to improve the overall efficiency of the work obviously must be implemented but they should be used only to complement the work undertaken by man.

Astronaut is forced to destroy the computer which decides he is not fit for the mission (*from* 2001: A Space Odyssey).

Living Underwater

The first problem to overcome in adapting to a sub-aquatic environment arises from the differences between air and water, from which all living organisms derive the oxygen that is vital to their existence. Since living matter originated in an aqueous environment, all animal life is presumed to have once been water-breathing. Air, however, seems to have several advantages over water as a respiratory medium.

The blood vessels of the human lung.

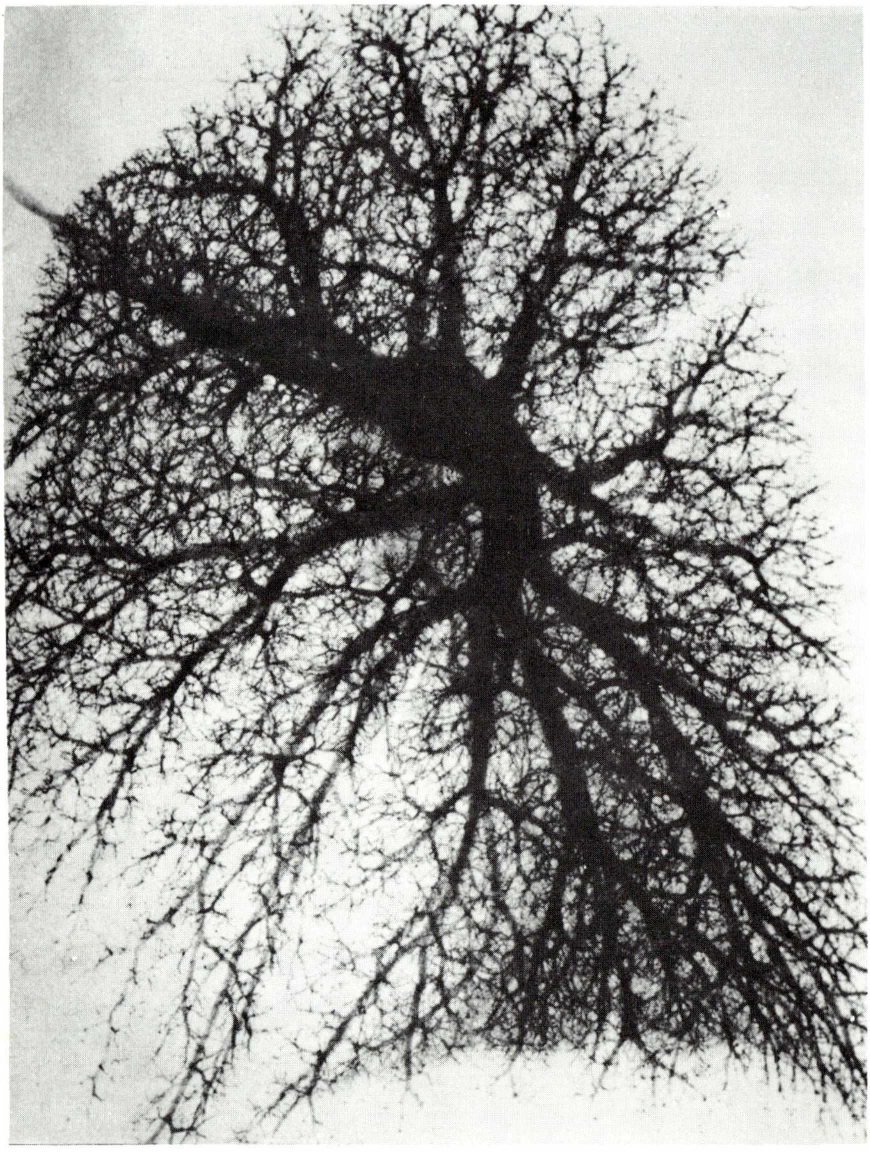

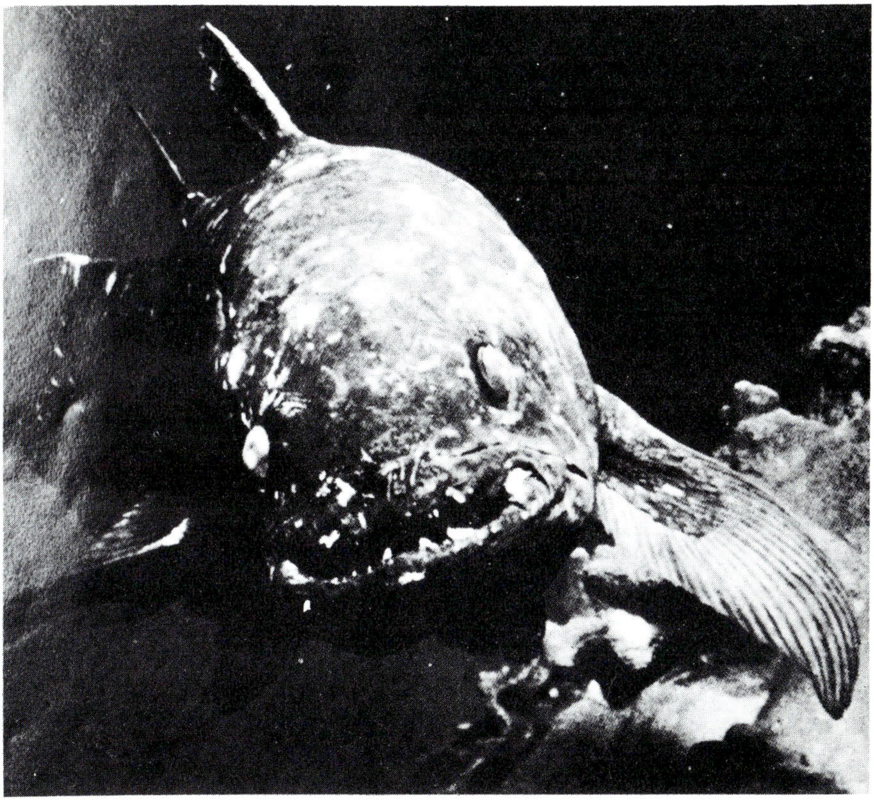

The living fossil coelocanthe laterimera *photographed live in the Indian ocean off Madagascar.*

In comparing air and water, a number of differences immediately become apparent. There is much less oxygen per unit volume in water than there is in air – a litre of air has enough oxygen to support a given animal 25 times longer than would be possible for an equal volume of water. Consequently, fishes must breathe a much greater volume of water in order to obtain the same volume of oxygen as an animal breathing air. The density and viscosity of water is much greater than air, and so requires more effort from the respiratory organism. The crucial difference, though, lies in the speed with which oxygen and carbon dioxide are exchanged through the semi-permeable membranes in the lung of an air-breathing creature and in the gills of a fish.

Most fishes have considerable difficulty surviving out of water. Apart from being unable to conserve body fluids in an essentially dry atmosphere, fish encounter serious respiratory difficulties. Even though the oxygen content of air is more than sufficient for their needs, they suffocate, because the delicate structure of their gills is mechanically unstable in air, and collapses, so that the remaining area is insufficient for gas exchange to take place.

Air breathing animals, if submerged in water, usually die from lack of oxygen,

An atomic model showing the random co-ordination of water molecules.

but for reasons that are not strictly comparable with the fishes' inability to cope with air. Apart from the fact that the oxygen requirements of most warm blooded animals are considerably higher than those of fishes, the composition of water inhaled by a drowning mammal most likely differs substantially from that of the blood; because of this difference a transfer of the solutes in the water and the blood takes place by osmosis across the semi-permeable membrane in the lung, fatally altering the composition of the body fluids and possibly damaging lung tissues.

Although there is a marked difference in the properties of air and water as respiratory media the process of gas exchange is similar in principle: the transfer of oxygen to the blood and excess carbon dioxide to the air or water across a semi-permeable membrane, in the gill or lung, gas exchange in air breathing animals is expressed in terms of simultaneous changes in the amounts of oxygen and carbon dioxide present within the organism. These are measured by the partial pressures of the gases within the blood. Gas exchange in the lungs and

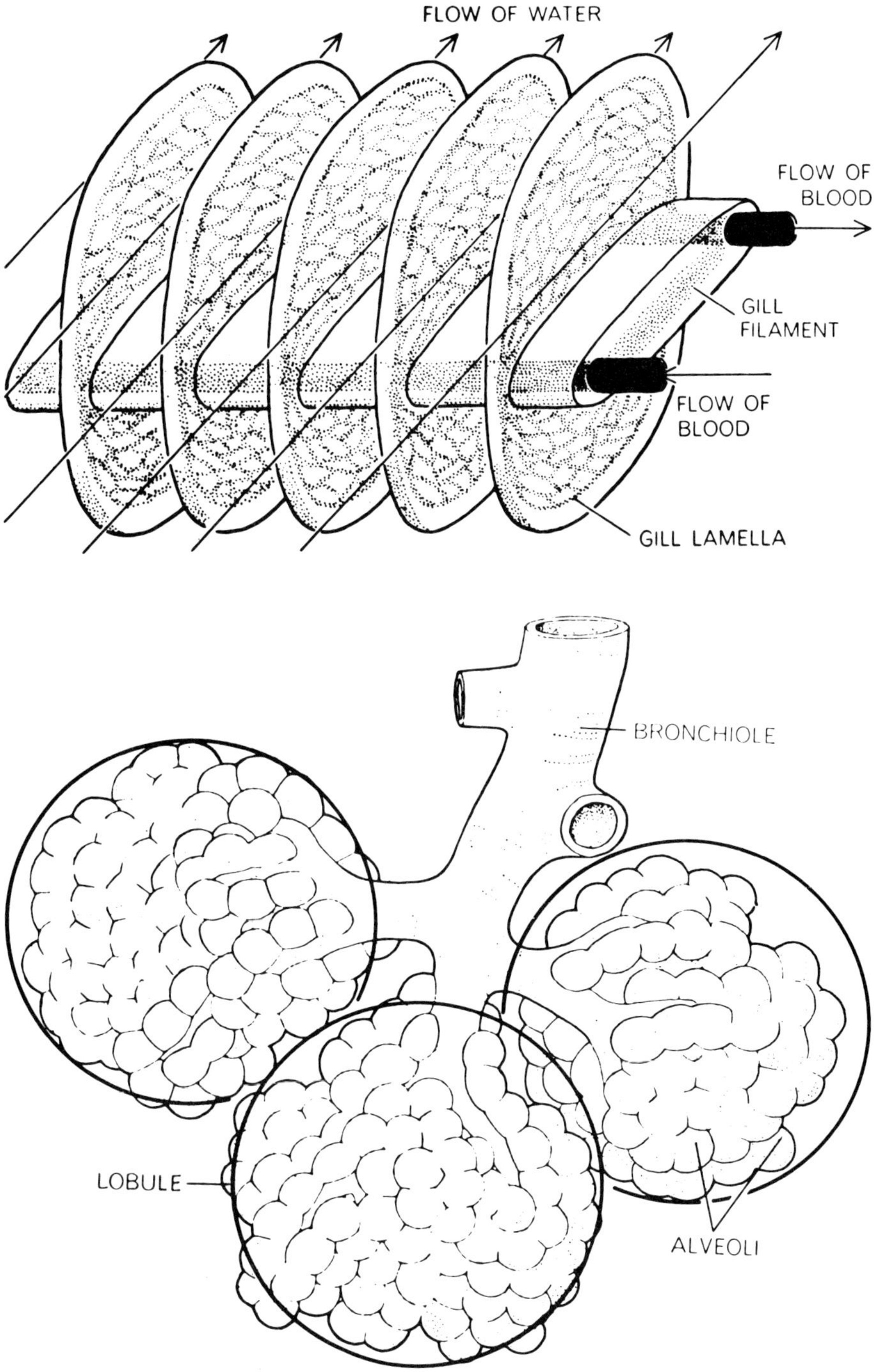

The structure of gill lamellæ (above) and lung alveoli.

blood also indicates the limitations imposed on aquatic respiration by various factors: metabolic rate, the ratio of gas exchange, the ventilation of the lungs, changes in the barometric pressure, the composition of the blood, and the composition of the inspired gas.

The shape and structure of the gill and lung both bear a complex interrelationship with the physical properties of their environment, and the gas exchange requirements of the fish or mammal. The significant difference between the gill and lung is related to the transfer of oxygen and carbon dioxide between the respiratory organism and the gas exchange medium. The effective solubilities of oxygen and carbon dioxide across a semi-permeable membrane in air or water (in which the respiratory gases diffuse about 6000 times more slowly), largely explains the differences in structure between the two organs.

Experiments carried out on liquid breathing dogs have enabled the determination of a mathematical lung model. Such a lung model turns out to be analagous to a sphere of liquid at whose interior surface gas exchange takes place. When a breath of air or water is taken into the gas exchange unit of the respiratory organism, the oxygen molecules must travel by diffusion through the gas exchange medium in order to be allowed across the semi-permeable membrane. Because this diffusion time is so fast in air, the fact that the oxygen molecules must travel a distance several times thicker than the alveolar membrane is of little consequence. The basic functional units of gas exchange in the lung are the primary lung lobules, each of which consists of a large number of alveoli where gas exchange actually takes place. The shape of the alveoli is near spherical and the size and number of the primary lung lobules compares with that of a hypothetical spherical lung of about 500,000 gas exchange units, 1 mm in diameter.

It is possible to develop by analogy a similar model for the gills of fish in which the primary gas exchange units are of a different structural form. If a fish were equipped with a hypothetical lung instead of gills it would virtually be deprived of oxygen in its blood because of the difference in shape of the gas exchange units. The spherical lung units are unsuitable for breathing water, whereas the flat plate-like lamellæ of the gill, with a continuous flow of water over them, overcome the diffusion problem faced by the spherical alveoli.

The comparative function of the lung and gill can be seen in terms of their operation as respiratory organs. In the lung, venous blood enters a random capillary network of the alveoli, and emerges oxygenated. Air enters and leaves the lung in a reciprocal fashion by means of an airway, so that the volume of air in the lungs from the gas exchange surface to the openings of the airway constitutes a deadspace. In the gill, however, there are two important differences, as the deoxygenated venous blood flows unidirectionally over the lamellæ, while the water stream is pumped over them in the opposite direction, providing a countercurrent system of exchange. Because the waterflow is unidirectional the gill does not have an equivalent deadspace. However, because of the longer diffusion times of oxygen and carbon dioxide in water compared with air, a diffusion deadspace must be taken into account. The large difference between the volumes handled in ventilation in air and in water required to maintain a particular oxygen level is due to the fact that only a relatively small amount of oxygen can be carried by water compared with air. The carbon dioxide levels

Snibby, the first dog to breathe water and live.

Skeleton of a whale showing extreme adaptation to aquatic existence.

Below right: A goat emerges from a pressure chamber after a simulated dive.

in fishes must consequently be kept low, as carbon dioxide has greater solubility in water than in air, and only a large gill ventilation reduces the level of carbon dioxide retained in the blood. Aquatic organisms are adapted to extremely low carbon dioxide levels in their blood and tissues as compared with air breathing animals, and their large ventilation of water, which is highly viscous, imposes severe limitations on their ability to function efficiently. Experiments have shown that in narcotised fishes, about 30 per cent of the resting metabolism is devoted to maintaining the work of normal gill ventilation; this rate can reach 70 per cent when more energy is being used. Also the ratio of gill ventilation to cardiac output of fishes is of the order of 15 : 20, compared with approximately 1 in air breathers. This large difference is attributed to the low solubility of oxygen in water, and raises the question of how much the ventilation of fish can be increased during work without becoming a limiting factor; it is possible that this difference may explain why the maximum oxygen absorption of fish does not exceed 4–6 times greater than its resting metabolism.

These considerations describe in a quantitative way the limitations of aquatic respiration, and point to the necessarily low carbon dioxide levels which must be present in the blood and tissues, and the relatively enormous ventilations which are required to achieve proper oxygen levels.

The physiological problems of men breathing hyperoxygenated liquids have

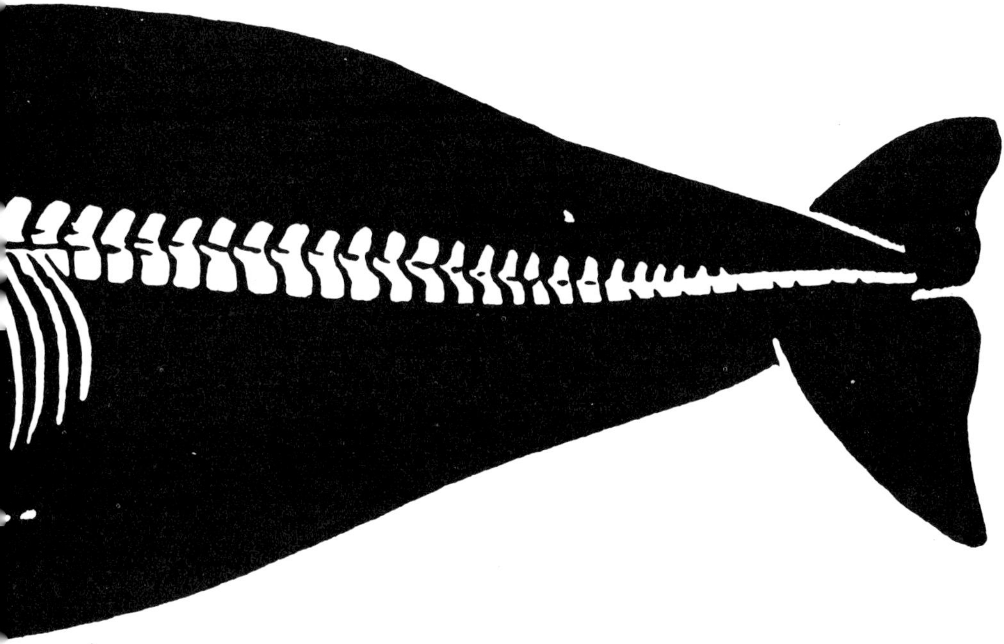

NOTICE
LEAKAWAY

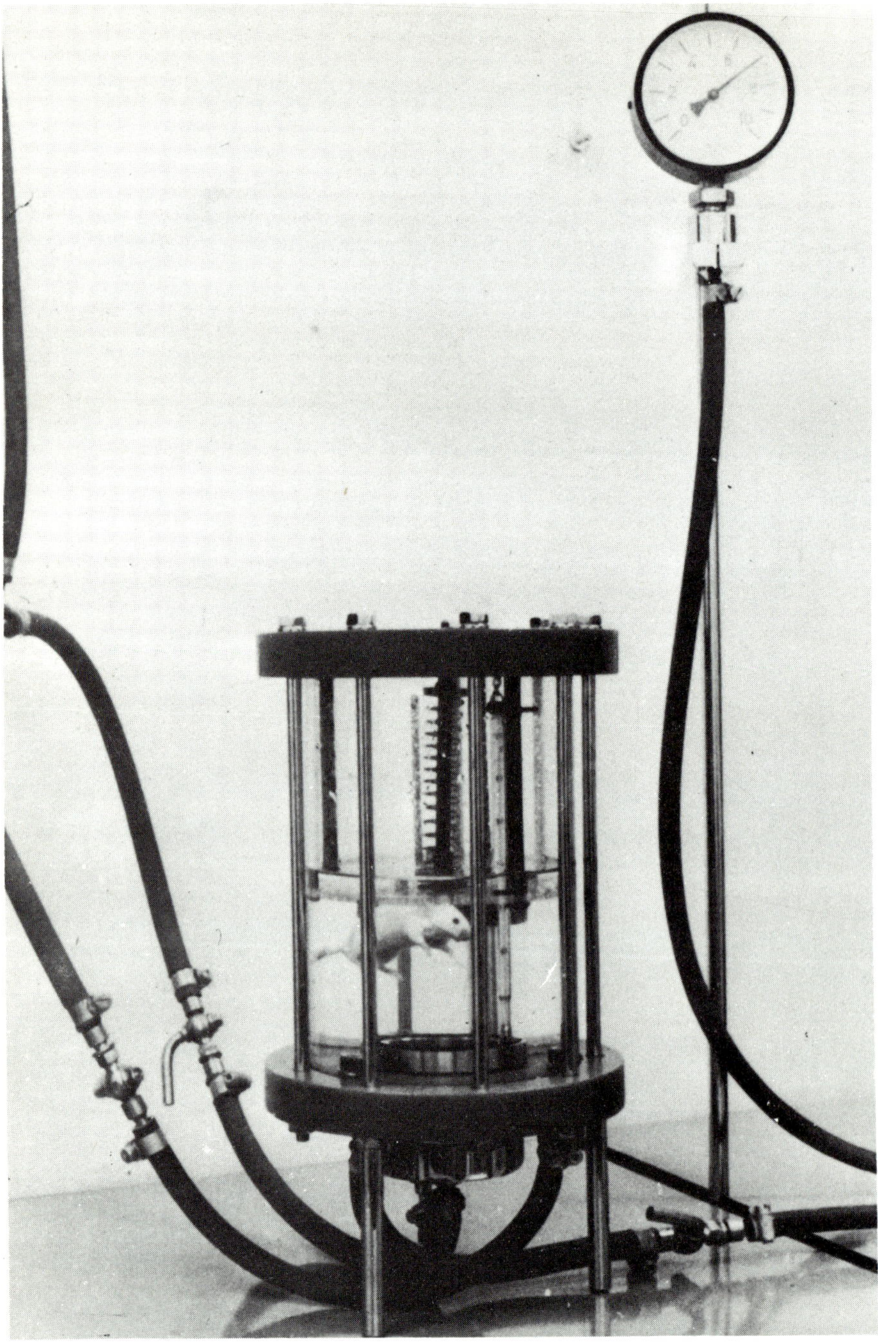

A mouse submerged in a chamber filled with heavily oxygenated liquid for a fluid breathing experiment.

A hamster in an underwater cage which is kept oxygenated by the water around it by means of a semi-permeable membrane that allows oxygen molecules in water through its 'pores'.

been studied by Dr Joannes Kylstra in the United States of America. Although man is certainly able to breathe liquid, there are problems to be overcome in order to make the system feasible for working divers. The first of these problems is that of keeping the arterial carbon dioxide levels within tolerable limits. An organic buffer, tris-hydroxymethyl-amino-methane (THAM) has been successfully used. This increases the effective solubility of carbon dioxide, which makes experimentally induced carbon dioxide retention tolerable over long periods of time. Certain disorders of the lung have involved irrigating one lung at a time with normal saline to 'wash out' the various secretions from the air sacs, and it appears that the flooding of one lung in conscious patients is quite tolerable. A diver volunteered to breathe water in one lung, while the other was supplied with oxygen. The lung was filled after the trachea had been anæsthetised. A liquid 'breath' of 500 ml of saline, at body temperature, was exchanged seven times. The diver, who was conscious throughout, felt no noticeable difference between the fluid-filled and gas-filled lungs. The only discomfort he experienced was when the tube was inserted into his trachea, and not during the flow of liquid in and out of his lung. There is no reason to suppose, therefore, that both lungs could not cope equally well when fluid-filled.

OXYGEN FROM SEAWATER

UNPRESSURIZED SYSTEM

P = 1 atm
150 mm Hg O_2
600 mm Hg N_2
10 mm Hg CO_2

PRESSURIZED SYSTEM

P = Ambient Seawater Pressure
150 mm Hg O_2
600 mm Hg N_2 (Equilibrium)
10 mm Hg CO_2
\+ Helium

The development of a liquid which has a composition comparable to blood plasma and is unlikely to damage the lung tissues, together with the necessary diffusion properties for the exchange of oxygen and carbon dioxide, would appear to be the solution to the problem. Synthetic fluorocarbons, which have

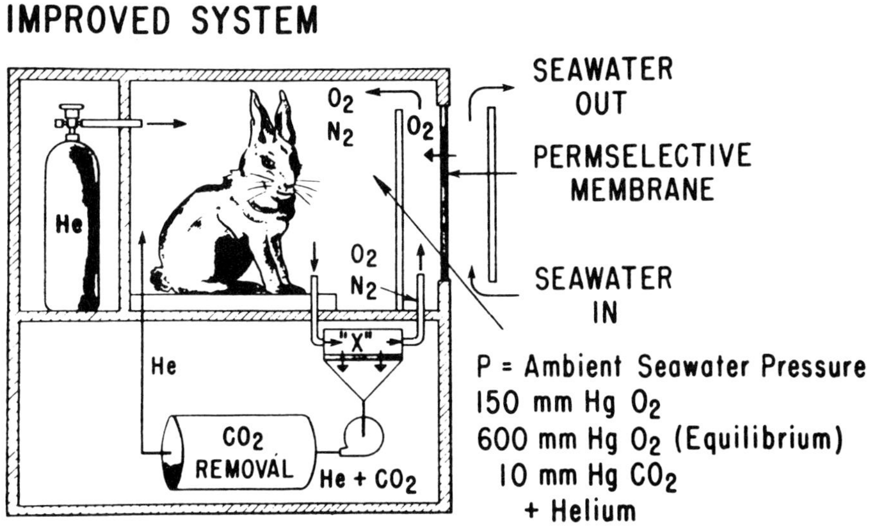

Oxygen from seawater.

much more appropriate oxygen and carbon dioxide solubilities for breathing than water, have very high viscosities compared with water and so are not entirely satisfactory. The extra work required for the respiration of fluorocarbons would invalidate the advantage in diffusion time. The effort required to breathe water, which is only 40 times more viscous than air, together with its slower diffusing rate of oxygen and carbon dioxide, remain the major setback to this system of fluid breathing. Mechanically pumping fluid in and out of the lungs presents a further difficulty: turbulence occurs in water at linear velocities 27 times smaller than in air. This greatly increases the pressure required for any given flow, and could possibly lead to the lungs collapsing during expiration of the volume of water that would be required for adequate carbon dioxide elimination.

The extra work required to breathe liquid compared with air is an obvious disadvantage. On the assumption that the diffusion times of oxygen and carbon dioxide were made tolerable with the use of a chemical buffer, or that a synthetic fluid with adequate oxygen and carbon dioxide solubility and with a suitable viscosity was used, the ability of the lungs to cope with the propulsion into and out of them will be the final deciding factor with this system of respiration.

The tidal method of respiration from the lungs complicates their ability to obtain oxygen from hyperoxygenated liquid. This is because the fluid which first enters the lung will have a closer proximity to the alveoli and its oxygen will diffuse into the blood stream. Fluid which is displaced nearer the centre of the lung and away from the alveoli will remain rich in oxygen owing to the slow diffusion time. This oxygen-rich water, however, will be the first quantity to

Snoopy.

Below:
An early undersea artist at work from within a rather rudimentary diving bell.

leave the lung because of the tidal system of respiration. Because the lung will always contain a volume of air or fluid even after it is fully expired, the fluid that is closest to the alveoli will quite likely remain there long after it has been deprived of its oxygen, and the carbon dioxide, also slow diffusing, will build up over time. Unless it becomes possible for a diver to breathe liquid at a uniform rate for a limited length of time the difficulties of this system of respiration might well be unassailable.

Assuming that the most significant limitations of breathing fluid through the lungs will be the relatively short time factor in which satisfactory gas exchange can take place, the possibility of an external artificial gill is worthy of consideration for longer dives. Such a synthetic respiratory system involves far greater technological problems than those that exist in terms of its physiological potential. These problems relate to the factors already described, which are concerned with the large difference in oxygen consumption for warm-blooded animals compared with fishes and the lower vertebrates, the specific differences between the lung and gill as gas exchange organisms, the physical properties of the gas exchange media, and the ambient environment.

Further, as the transfer of solutes across a semi-permeable membrane is primarily regulated by the process of osmosis, this system is not completely sufficient for biological purposes, and an additional system must actively assist the transfer of particular solutes to the cells of the organism, and the expulsion of specific metabolic products. The presence of electrolytes in biological fluids provides this additional system of 'active transport' by means of a charge carried or developed by the membrane, which makes it variably selective to the ions it accepts or rejects. Most bony fishes have a lower osmotic concentration than the surrounding sea water, and would literally dehydrate without the gill's capacity to return the excess salts obtained by active transport into the water. The development of a synthetic semi-permeable membrane for an artificial gill must be able to exercise similar control to 'active transport' in order to maintain normal blood composition. If such a membrane were developed for the artificial gill, it might also be possible for it to cope with, wholly or in part, the function of the kidneys.

If an artificial gill is used, normal respiration by means of the lungs must be isolated. Surgically, a cardio-pulmonary by-pass of the main blood supply to

An octopus.

Left:
Captain Nemo looks at a giant squid through the window of his submarine Nautilus (*from* 20,000 Leagues under the Sea, *by Jules Verne*).

the lungs could be made close to the heart, and the blood diverted to the gill system by grafting connections to the major blood vessels concerned, taking these to the surface of the body, probably at the top of the chest. For limited depths, the lungs may be inflated with inert gas which would enable normal speech to be carried out. At greater depths, however, the lungs would need to be flooded with a balanced saline solution. It would be impossible to keep them inflated with gas at extremely high ambient pressures, since it would diffuse rapidly into the blood stream through the lung membrane, making it necessary for the diver to carry large quantities of gas to meet this consumption.

The mechanics of the hypothetical gill system may have to assist the heart with the circulation of blood, because in warm-blooded animals the ratio of gill/lung ventilation to cardiac output is much lower than in fishes, mainly because of the low solubility of oxygen in water. The blood supply passing through the gill system would clot unless an anti-coagulent is maintained in the blood stream in adequate proportion. However, this is a great limitation as anti-coagulents cannot be endured for prolonged periods of time without having adverse effects. The heat lost over the large surface area required for efficient gas exchange would need to be controlled and insulated as far as possible in order to maintain normal body temperature. The change in solubilities of oxygen and carbon dioxide with salinity and/or temperature of the water will naturally increase the gill ventilation requirements. At great depths, the availability of oxygen rapidly depreciates, to which the natural gill automatically adapts. The mechanics of the gill system will have to be sensitive to variation in the oxygen content of the water and be able to adjust accordingly. The problem of a variable solubility in this context is important, considering that there is less oxygen than is needed to meet most mammalian requirements. The change in solubility for carbon dioxide is unlikely to hold any significance, since the enormous difference in comparative volumes between the blood and water would nearly always enable adequate diffusion to take place.

Alternative respiratory systems enabling average ocean floor depths to be reached, where the pressure is equivalent to 400 times that on the surface, would dramatically alter the respiratory requirements. They would probably also necessitate changes in the general metabolism of the diver. It is unlikely that either lung or gill respiration would be able to cope with the minute exchange of volumes of oxygen and carbon dioxide required at these depths. It is possible that some means of carefully controlled diffusion of oxygen into the blood stream, and carbon dioxide out of it, may be found with a flooded lung system which does not use the ambient environment as the respiratory medium, but carries hyper-cooled supplies of oxygen and inert gas, which are fed in closed circuit into a suitable respiratory fluid.

The River of Life, by William Blake.

The personalities of the individuals involved on each underwater mission, and their compatibility as a group, have to be considered in the design of every unit of submergence hardware. It is also important to develop selection criteria related to the physiological and psychological tolerance of the individual in specific environmental conditions. To help in assessing the design constraints in particular circumstances, operational personnel can be divided into separate groups, according to the kind of work to be undertaken and to the anticipated stresses.

Since no two individuals will be biologically identical, responses to stress vary, and the range of adaptation to stressful conditions is unique for each person. Some conditions of atmosphere contamination and certain psycho-physiological stresses may lead to diseased conditions or pathological changes without necessarily harming the performance of the persons involved. For this reason criteria for the selection of personnel need to be flexible enough to adapt to individual variation and particular circumstances. The factors to be considered when classifying different underwater working groups are:

1) The barometric pressure of the ambient environment, and the related problems associated with prolonged exposure to high ambient pressure.

2) The atmosphere including the effects of breathing air and gas mixtures at high pressures and the associated complications such as the toxicity of oxygen, the narcotic effect of inert gases, problems of pulmonary ventilation.

3) The temperature and thermal conductivity of the ambient environment: the temperature of the water is an important factor for oceanauts who must be shielded from heat loss in order to perform useful work for any length of time. The high thermal conductivity of synthetic atmospheres, such as helium and oxygen (HeO_2), raises the room temperature by several times according to the atmospheric pressure. Also, isothermic clothing is required to maintain normal body temperature.

4) The duration of the dive: the time to be spent under water, whether confined within a submersible or isolated in an undersea house, is a determining factor in the selection of the crew. Some divers may be able to perform quite satisfactorily for relatively brief periods of time, but a different aptitude and depth of motivation may be needed by those who are prepared to remain under water for prolonged periods, or perhaps indefinitely.

5) The type of mission: the selection of operational personnel for a specified mission, their individual degrees of tolerance, group compatibility and the imposition of particular stresses expected to be encountered, are all crucial design considerations.

6) The psychological effects of transplantation from a natural environment into the synthetic one of the submergence system and an alien one under water must also be taken into account.

From these criteria it is possible to indicate three basic group classifications, although the problems are nearly always inter-related with those of another group. Two groups refer to the diver:

a) *The surface diver:* who is under water for specifically limited and relatively short periods of time.

b) *The saturation diver:* who must spend long periods under water and make a necessary transition from normal terrestrial life to a completely synthetic

U.S.NAVY
DSRV-1

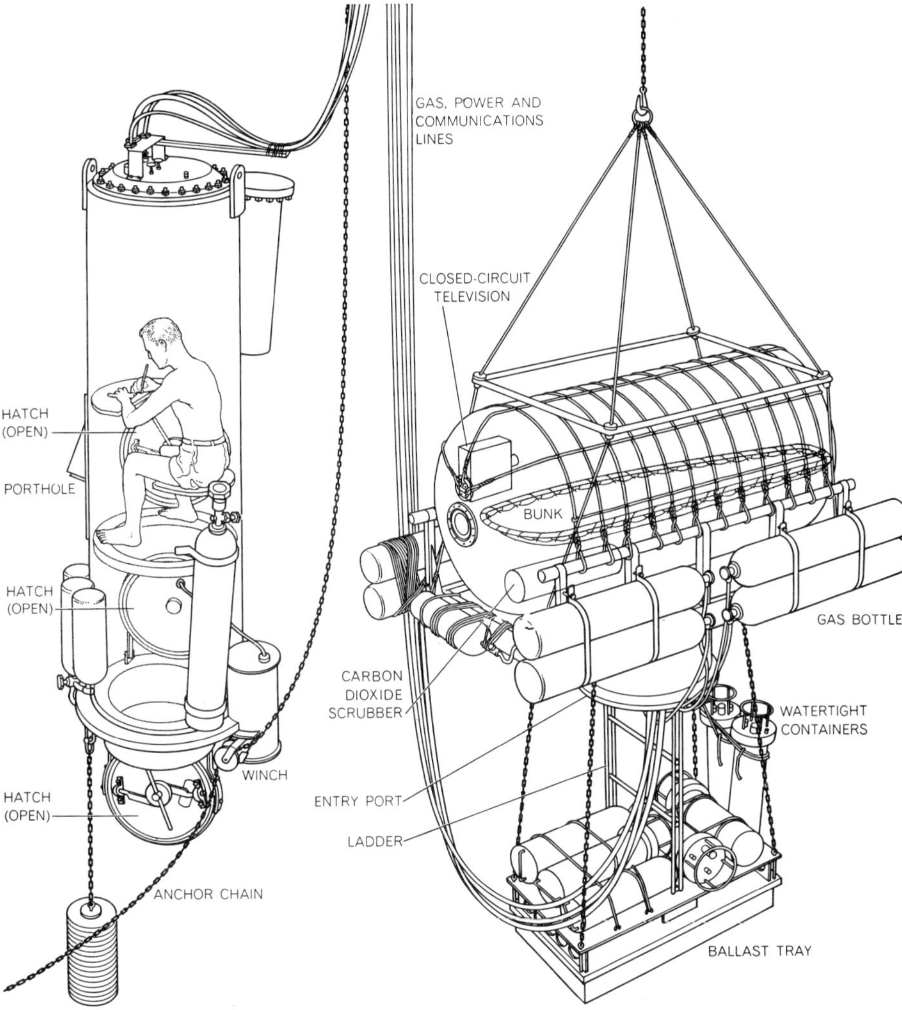

Above:
Two chambers used in the Man in Sea 432 ft two-day dive. The 'submersible decompression chamber', or SDC (left), is an aluminium cylinder 11 ft long and 3 ft in diameter. With the hatches open and the inside gas pressure equal to the external water pressure, the SDC serves as a diving bell. The SPID (right) is an 8 by 4 ft inflatable rubber dwelling with a steel frame and ballast tray. Access to it is through an open entry port at the bottom.

Left:
Lockheed deep search and recovery vessel simulator and pressure hull model. This search and recovery system is being developed for submarine rescue but the technical spin-off will have an important influence on submersible guidance in the future.

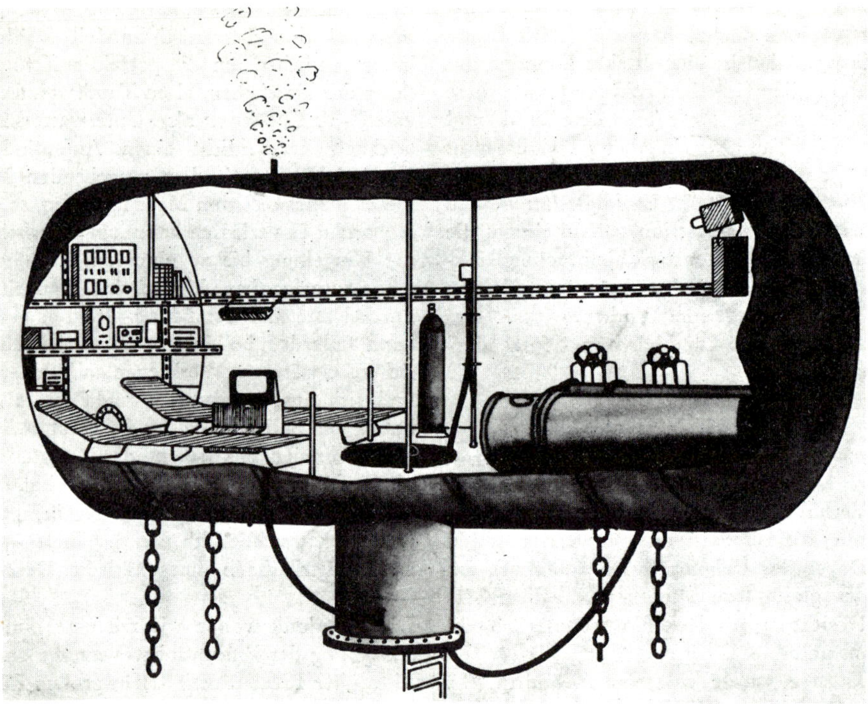

Precontinent 1.

behaviour pattern developed for the particular engineering techniques, with the operation of which he is involved under water.

c) The third group is predominantly made up of the crews of submersibles who must remain confined in a restricted space for specific but sometimes prolonged periods of time. This group may include members from both groups of diver, since the surface diver may well be operating from a submersible chamber in which he may be confined for extended periods of decompression. Equally, the saturation diver must spend his period of decompression (usually at the surface) within the restricted space of a decompression complex. If craft are developed with hyperbaric chambers, in which divers have to be kept saturated for a particular depth, this may also entail long periods of time spent in a restricted space. Within these groups the selection of personnel may involve specialist individuals whose capacity to undergo particular stresses may not be compatible with the strong motivation of an experimental subject.

Anticipation of the physiological, psychological and environmental stresses on man, and the physical stresses on the operational hardware, leads inevitably to the adoption of a design method where the number and complexity of the systems and components is deliberately limited.

The minimal system approach offers a number of distinct advantages in supporting life, as it tends to reduce substantially the possibility of life-endangering failures, either mechanical or human. Any failure within the life-

Friendly Samantha, the wild sea-lion; Sealab 2.

The Aluminaut *was launched at the Electric Boat Division of the General Dynamics Corporation in September 1964.*

During the sea trial period, a highly successful long endurance submergence was conducted by crossing the Straits of Florida from the Bahamas to Miami. The vessel stayed submerged for approximately 33 hours while travelling at an average depth of 1,250 ft. During the transit, the Aluminaut *covered a distance of approximately 70 miles at speeds up to 3.5 knots and descended to the bottom of the Straits to 2,750 ft. The* Aluminaut *surfaced with approximately 12 per cent of her battery capacity and 24 hours of life support remaining. During the dive, stress information on the 6.5 in. thick aluminium hull was measured.*

During sea trials, Aluminaut *demonstrated the feasibility of controlled drift in the Gulf Stream for scientific observations on a 25 mile 'free ride' at depths in excess of 1,000 ft and in currents up to 3½ knots velocity.*

During sonar trials, several sunken vessels were detected off Fort Lauderdale, Florida. A large World War I concrete barge section was located that appeared to be 80 per cent buried in sediments evidently deposited since that time. Aluminaut *sonars include a continuous-transmission, frequency-modulated sonar used primarily for bottom navigation; a high frequency sideways-looking sonar which provides a high resolution permanent shadow-graph record; and a utility sonar which can be mechanically trained by a pan and tilt mechanism on the bow. Other echo sounders are installed for determining distances to the surface and the bottom.*

In November a deep diving trial was conducted some 20 miles west of Great Abaco Island in the Bahamas to a new submarine depth record of 6,250 ft, and allowed a systematic collection of stress data from external and internal strain gauge systems.

After the 6,250 ft dive, successful completion of all trials indicated that Aluminaut *was ready to provide charter services to the scientific, defence and industrial communities.*

Under contract to Special Projects Office in January 1966, Aluminaut *ran tests at depths to 2,500 ft on the Deep Submergence Systems Group's Doppler navigation equipment which was to be installed on the bathyscaphe* Trieste. *During these tests, 50 ft of aluminium pipe were laid by the* Aluminaut *on the bottom of the Straits of Florida in precise alignment with the current in order to permit exact calibration tests of the Doppler system.*

Aluminaut *was called upon to take part in Aircraft Salvage Operations, Mediterranean, off Palomares, Spain, when a B52 bomber and the KC135 tanker aircraft collided. Approximately 30 dives were conducted in the search for the missing H-bomb, averaging 2,300 ft and well over 7 hours in duration. Six crewmen were used on practically all dives to permit a comprehensive sonar as well as visual search. The deepest submergence was 3,000 ft where the bomb had lodged after the first recovery attempt.* Aluminaut *conducted the first successful rendezvous with another deep submersible, the* Alvin, *on 15 March. At this time,* Aluminaut *was directed to bottom in close proximity to the weapon, and maintained a watch for almost 24 hours. This operation was repeated when the weapon was relocated on 2 April.*

Although Aluminaut *was not called upon by the Navy to retrieve the weapon, a droppable grapnel and snap hook system had been developed which would have permitted this achievement.* Aluminaut's *reserve buoyancy at the depth of 2,800 ft is approximately 5,000 pounds. This buoyancy is created by dropping steel shot. To provide redundancy,* Aluminaut's *high pressure air system is able to provide an additional 3,000 pounds buoyancy.*

Many different types of biological life were observed during the search. During all of the submergences, only one sea creature was observed to dart away from the Aluminaut *in fright.*

The Aluminaut *manipulator device (AMD) – two 9 ft articulated arms capable of lifting 200 pounds each – was installed in July 1966. A 7 ft external stowage compartment for specimens, installed near the forward viewing ports, holds approximately 600 pounds. Arm-mounted racks held biological sediment, coring, and other recovery tools. Removable aircraft-type wheels were also installed for manoeuvres in close proximity to the bottom.*

In August-September 1966, a company-sponsored resource cruise was made on the Blake Plateau where a vast company of manganese oxide was found and explored.

In October-November 1966, while working under contract for the US Naval Oceanographic Office. Aluminaut *performed a unique ocean floor search and recovery of instruments in a depth of 3,150 ft of water. The operation was similar to the one carried out to recover the lost H-bomb off Spain.*

The search for the link of oceanographic current meters which was 2,900 ft long was considered to be the most difficult part of the mission. On the first dive Aluminaut *took 1 hour 56 minutes to locate the instruments which had been lost for over a year. On the second dive,* Aluminaut *was unable to re-locate the equipment for 3 hours 51 minutes, because it had come across another cable which had first to be identified,*

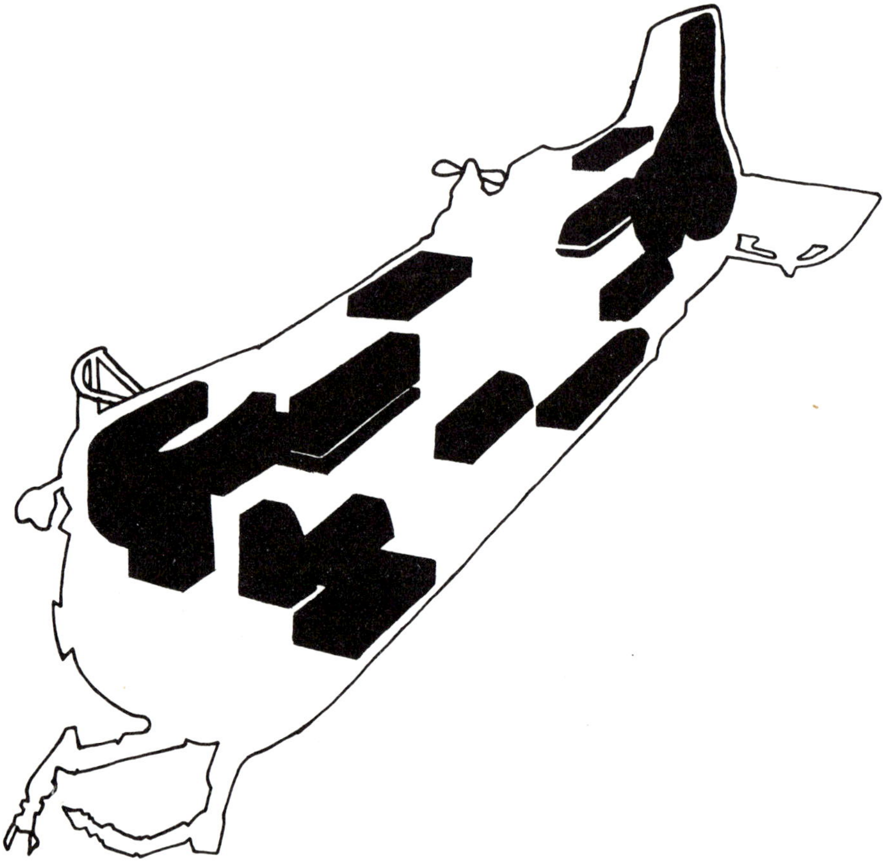

and was accordingly examined by the scientists on board, using the two mechanical arms of the submarine to lift the cable in front of the portholes for viewing. With a special spool device and nylon and steel cables, recovery of the instruments was effected successfully without damage to the meters. Great care was required in manoeuvring the Aluminaut *at depths where visibility is negligible, so as to avoid becoming entangled in the various cables and lines used for recovery or lying on the ocean-floor. Once the recovery cables were attached to the instruments, the* Aluminaut *was able to resurface at speeds of up to 150 ft per minute.*

support systems can present a direct threat to life and in some cases warning is given only minutes before disaster.

Provision must be made for an emergency escape from an underwater craft in the event of an unpredictable accident. Saturation divers who are forced to surface under such circumstances face serious decompression problems. A combination of this and other factors can produce severe physiological stresses in those involved.

A decision to adopt a minimal system design, where the safety of the operational personnel is the paramount consideration, will lead to a significant point in the determination of safety criteria. Design approaches which take into account only purely physiological and engineering factors are simply not adequate, regardless of how well the operational personnel may be motivated, or the extent to which their adaptive capability, and general tolerance, is developed. Experiments where design criteria have been solely related to physiological safety factors have also incurred difficulties and accidents.

It is vital to determine design constraints by an integration of the physiological, psychological and environmental factors involved, so that there is a complete understanding of the overall problems. Lack of co-ordination and understanding between the different disciplines may lead, literally, to disaster. Design criteria for submergence hardware must have an inherent flexibility that

Inside Sealab 2, *an aquanaut fills out one of the log books.*

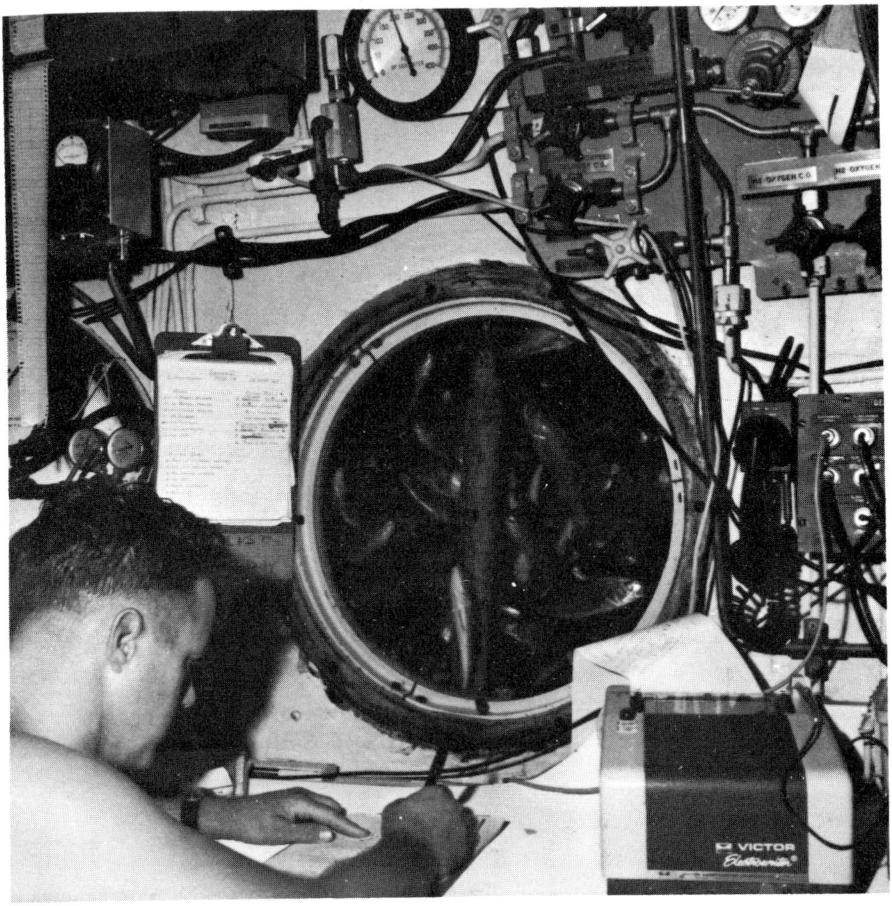

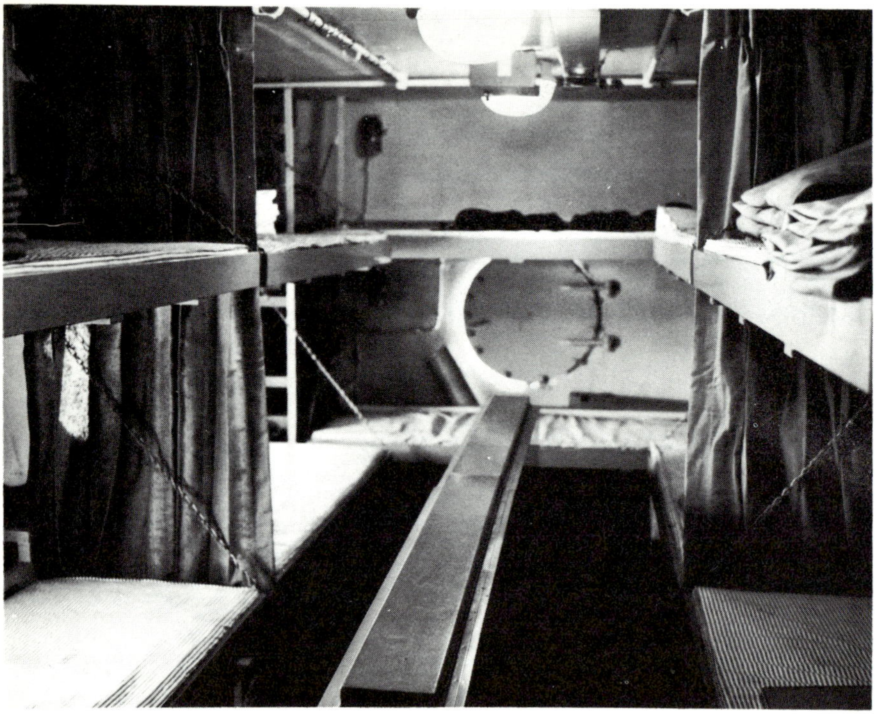

Sealab 2: *the sleeping quarters.*

would enable each project team to manipulate them as appropriate to the particular mission, circumstances and individuals who will operate within the system. There are two categories in which such criteria may be developed: the first concerns the tolerance of the individual and the group; the second, the technical feasibility of the mission.

For oceanauts the primary selection criterion is concerned with how well the individuals apply themselves to training and general 'dogsbody' work related to particular experiments. The selection of personnel for the *Precontinent* experiments (see Appendix 2) has verified this. It was decided for the first two experiments with the *Precontinent* that the individuals selected to participate should not be considered as record breakers or as exceptional divers, if the experiments were to produce a wide range of results with practical applications. The distinguishing feature of the oceanauts chosen was their intense group loyalty throughout the experiments.

Individual differences in certain psycho-physiological processes, particularly those known to be important components of intense emotional response, are significantly related to adaptation to stress. There are reliable links between certain responses of the body to stress and observable personality traits, and a degree of displacement and recoverability of character. Blood pressure, heart and respiration rates also give some indication of reaction to stress. In *Sealab 2* and *Precontinent 3* (see Appendices), use was made of electroencephalogram

Sealab 3: *aquanauts at work.*

(EEG) and electrocardiogram (ECG) telemetered information from divers carrying out various work tests in order to evaluate those responses which might indicate the extent of adaptation and work performance under water.

It would be unnatural or unrealistic to assume that freedom from emotional crises or psychological upsets could be maintained in stressful and alien environments. Technically competent individuals may show varying degrees of psycho-pathology during prolonged periods of confined or isolated work underwater and still be effective both in their jobs and as members of a small interdependent group, or operational crew. Often obviously neurotic mechanisms, even though they may border on the pathological, need not be unacceptable. In the experience of Commdt. Cousteau, 'an individual having a difficult character under normal conditions may well become more difficult in trying times or circumstances, or on the contrary he may, by necessity, become an excellent and dependable comrade.' Greater emphasis should be placed on the adequacy of psychological defence mechanisms rather than upon the nature of

Sealab 2: *the habitat.*

them. However, persons whose past suggests an emotionally unstable character must be considered unacceptable.

Motivation plays an important part in adaptation to undersea environments. It is essential that all categories of personnel working on undersea experiments should be volunteers. The reasons for an interest in undersea exploration are often different for the amateur and the professional, but even with the latter group, material benefits or money are rarely the main motivating factor.

Mission success depends on more than the individual's efficiency. When a small group of highly skilled and trained men are forced to work in isolation for a prolonged period, mission effectiveness will suffer unless the group is exceptionally compatible. With the passage of time under conditions of extreme isolation and confinement, emotions distort both reason and judgement, and a situation may develop where there is constant inter-person friction.

All undersea work requires the integration of the individual within a group, where he must hold responsibility equally to his team and mission accomplishment. Most friction arises when the individual feels he is imposed upon by the group, and so the design of the undersea habitat should provide an opportunity for privacy when it is required.

The nature of work under water, the length of dives, and the temperature of the water greatly influence food intake, and the provision of an ideal diet is a complex problem. In a synthetic atmosphere, it is very difficult to cook anything hot because, with the high pressure and the thermal conductivity of helium, water boils at 160°C. Relative humidity is 100 per cent, and further complicates eating habits.

Sealab 2: *a pressure-resistant container used to carry food and equipment to and from the habitat.*

The factors that normally determine food requirements are related to health, metabolic rate, and the activity of the individual. The calories required for various tasks have to be assessed with the consideration that food conversion is dependent on the oxygen available in the atmosphere. It is possible that specific diets could be developed to extend the tolerance and performance of operational personnel.

The *Precontinent* experiments (see Appendix 2) have provided valuable indications on the dietetic problems of oceanauts. During the first two experiments (1962 and 1963), gastric discomfort occurred, due to the provision of starchy foods and dishes with sauces. The toxic effect of alcohol, as well as problems of atmosphere contamination with aromatic foods at increasing depths, were noted. In the diet for *Precontinent 3* (1965), the calorie value for the men working at 13°C was 4000. The food balance was 40 per cent glucose and 20 per cent protein. In order to prevent atmosphere contamination, all the food was frozen

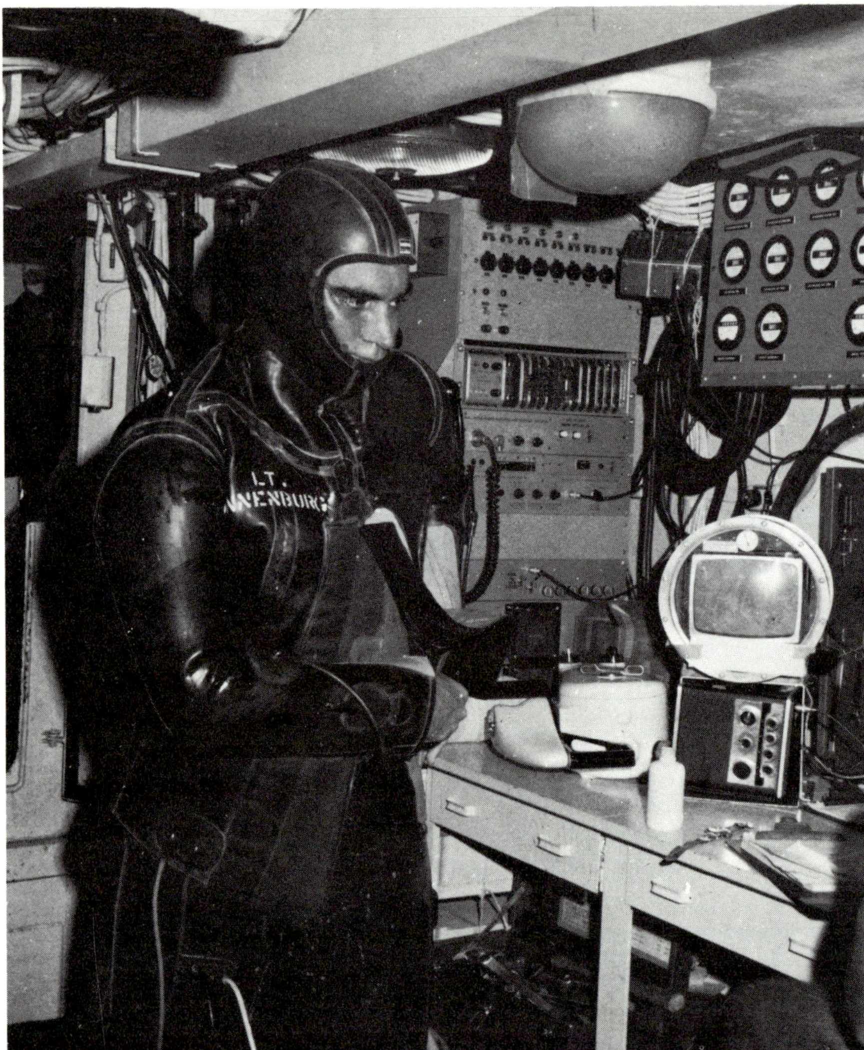

Inside Sealab 2.

or tinned, and re-heated in an infra-red oven when required. The food was grilled or poached to avoid gastric illness, and all sauces and dishes that could have caused fermentation were systematically eliminated.

Familiarity with operational conditions before an actual mission is an important psychological advantage. Experience of a task even under simulated conditions makes the real job less formidable, and is especially important to men facing prolonged immersion.

Because of the absence of adequate sunlight at depth, a new diurnal cycle may be developed, either through physiological need or in order to improve

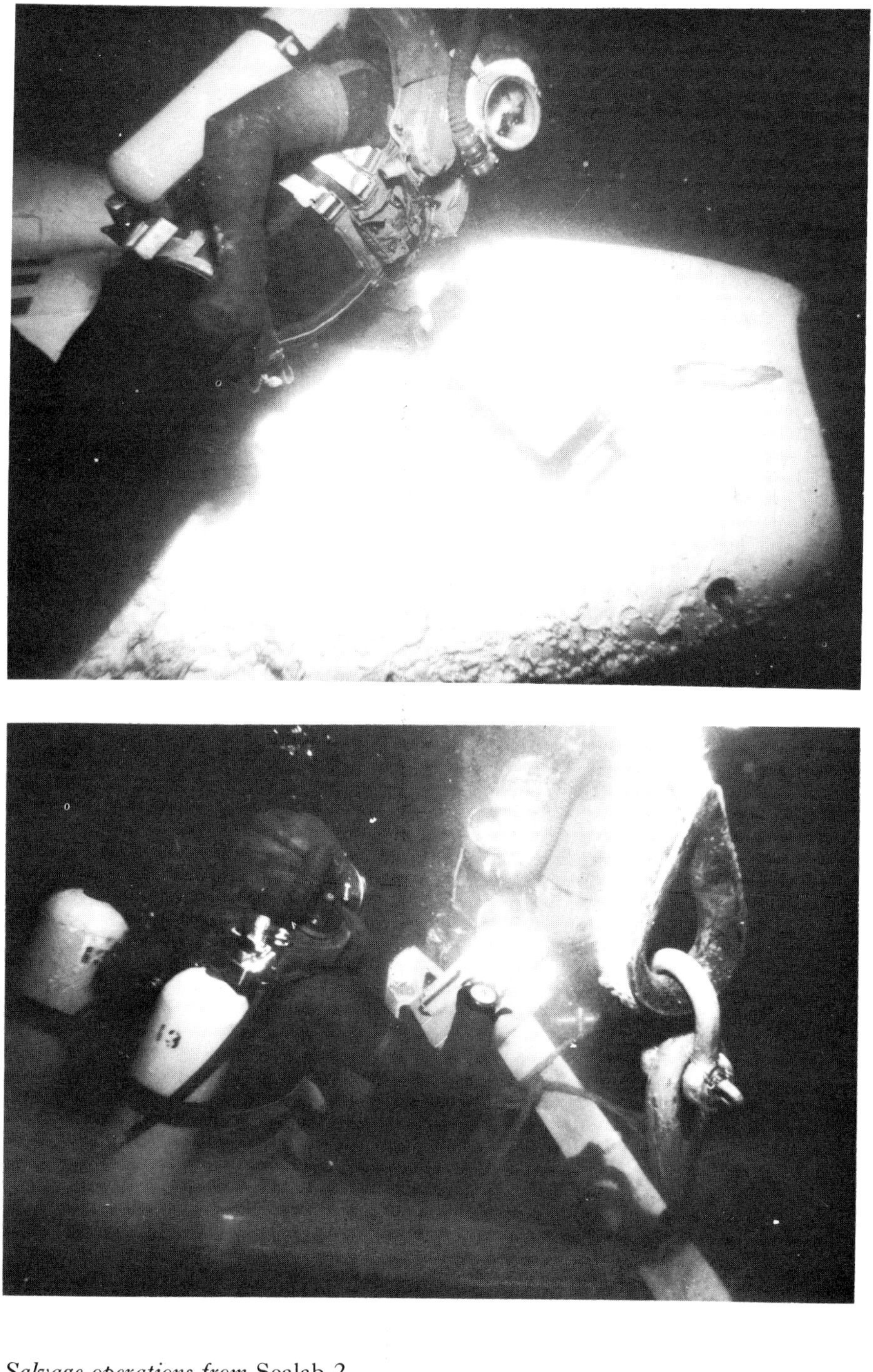

Salvage operations from Sealab 2.

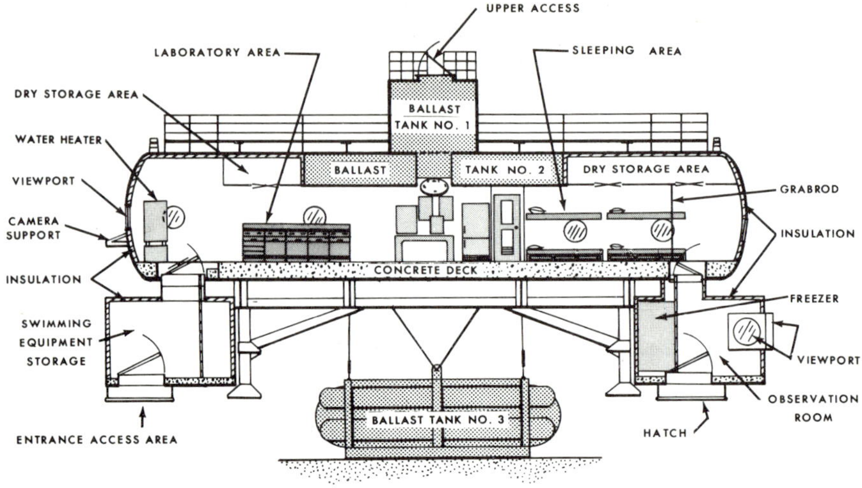

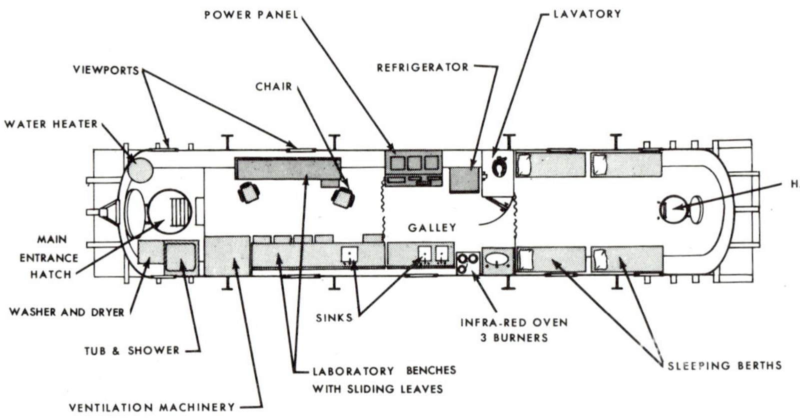

Above: Sealab 3: *interior, cross section.*
Below: Sealab 3: *interior, top view.*

the efficiency and comfort of the crew. Lack of sunlight also means that ocean-auts must receive ultra-violet radiation from an artificial source.

Extension or adaptation of the sensory systems may become necessary, and can be achieved by activating the neurological system with suitable trigger stimuli. Sensory thresholds can be altered by preliminary 'get-set' signals. During World War II, the use of red lighting in ready-rooms and dark-adaptation goggles established the importance of exposure to stimuli before a critical judgement was required. The ultimate technique is that of not merely alerting a total system for any signal or physiological response, but enhancing its level of performance in relation to a specific stimulus and establishing a selective filter in that system. In this way, a man could be alerted to a danger

Sealab 3: *The modified* Sealab 2 *with two extra cabins.*

signal, and react practically to it, without consciously recognising it as a hazard, and without undue emotional stress.

It is also possible to improve the performance of operational personnel with the use of drugs. The effects of this may range from direct contributions in energy conversion, improved attitude, or simply a feeling of competence. In addition to these effects upon the supportive physiological mechanisms are those related to the neurological aspects of saturation upon which human performance depends: perception; mental processes; motor activity; communication. However, there is a real need for a transition from what has been done

Sealab 3: *inside the systems are more sophisticated than* Sealab 2, *but the design method much the same.*

Right:

Sealab 3: *the aquanaut suit has a radioactive isotope which heats the water round the diver's body; the helmet is fitted with diver-to-diver communication via an unscrambling unit on the surface which minimises distortion of the voice when oxy-helium gas mixtures are being breathed.*

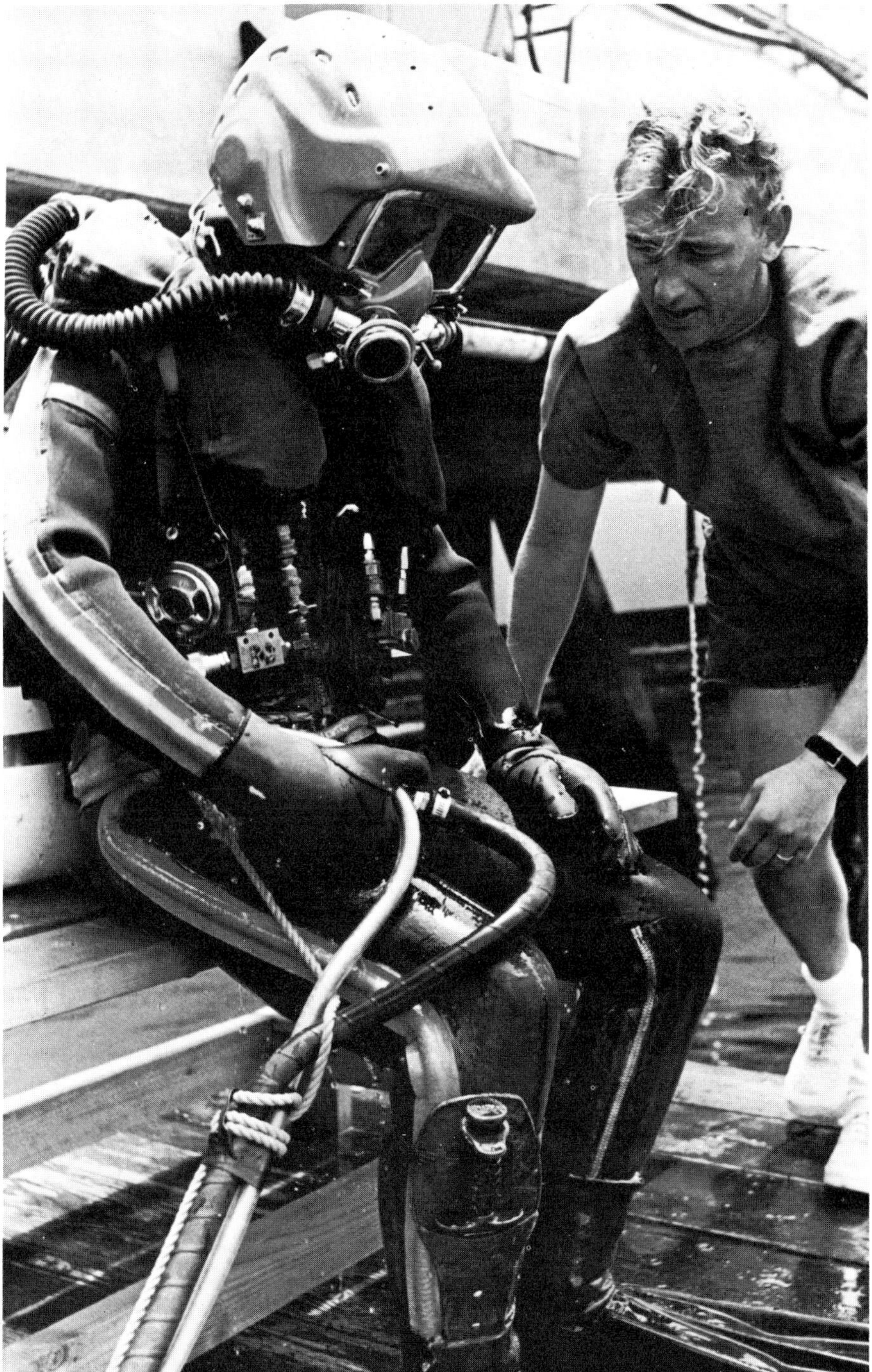

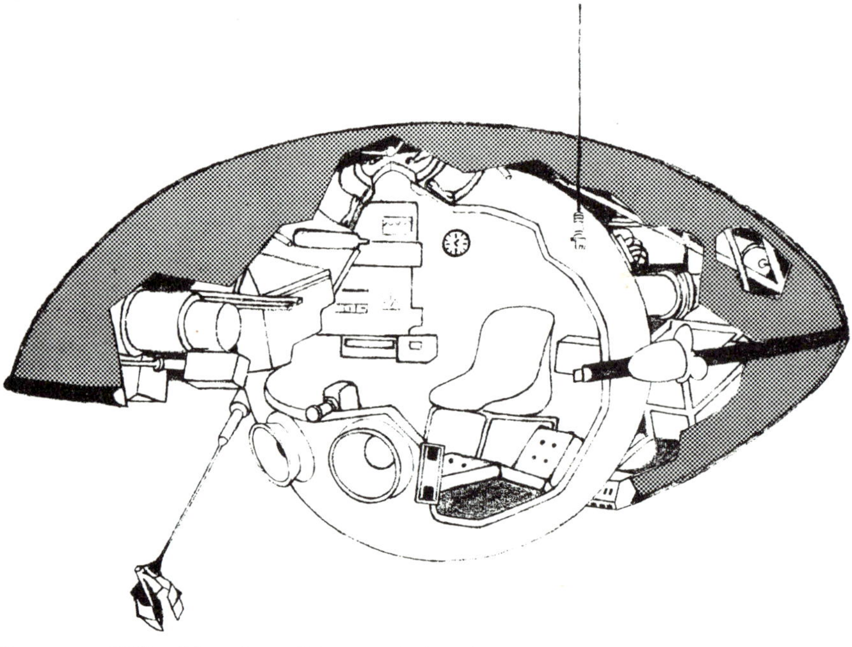

Deepstar 4000: *this submersible was designed and built for Westinghouse Electric Corporation by the Centre d'Etudes Marine Avancées. The number '4000' denotes its ability to operate at a depth of 4,000 ft. It weighs 7.9 tons and measures 5.7 m long by 3.4 m wide and 1 m high. The spherical pressure hull takes three crew and the craft has a cruising speed of 3 knots from two 4cv motors, together with a self-contained electrical and breathing system.*

Sealab divers and photographers feeding Tuffy the porpoise in feeding pen.

or observed to what can be deliberately brought about by skilled use of psychopharmacology. So far it is possible to overcome the feeling of fatigue, to reduce anxiety and the feeling of threat, to extend physical and mental performance, to influence perception of time and consciousness of others, to heighten sensory awareness, and to influence mood, all of which factors are basic components of group and individual performance.

It is necessary to account for fatigue of the personnel, particularly those exposed to saturation dives. Fatigue is a response to a variety of factors, depending on the duration of exposure to fatigue-producing, elemental, and environmental stresses.

When the level of participation varies, the basic state of physical and emotional health, the manner of working and living, fatigue or the absence of it, are exhibited to varying degrees. Some of the factors which play a part in producing fatigue are: ventilatory effort in breathing compressed air or gas mixtures in hyperbaric conditions; repeated exposure to cold water; the 'loneliness' of the depths, or the effort of swimming in currents and turbulent waters; long hours of work with irregular meals; and lack of adequate rest, depending on the circumstances encountered. Although it has been noted that the fatigue experienced in saturation dives is usually excessive, it is not necessary to avoid fatigue altogether in work under water.

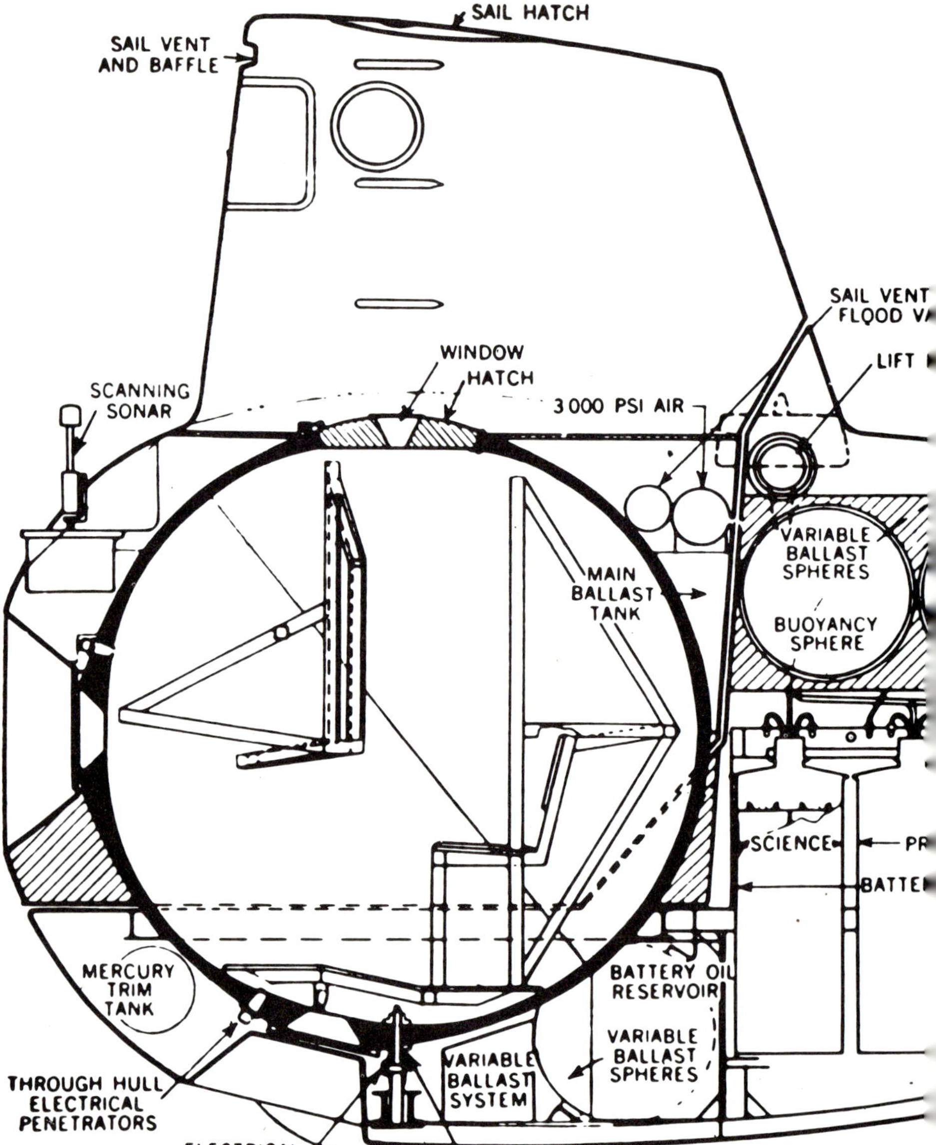

Cross-section of research submersible Alvin.

It can be seen that there are many interrelated factors involved with maintaining a human being under water. A successful integration of these different problems will enable man to survive and work usefully even though it is inevitable that the stresses encountered will impose new patterns of behaviour. The problems of communication related to helium speech distortion, and

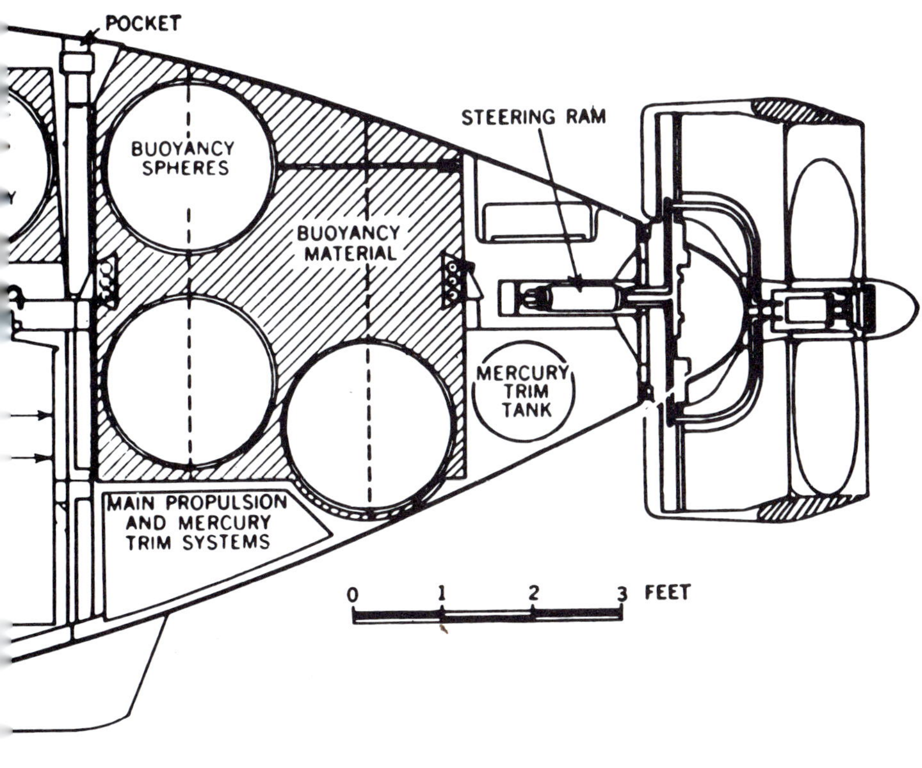

equally the general lack of interest in 'the surface world' that develops during saturation dives must be accounted for in the design of operational equipment. Such problems are mainly related to inadequate and impersonal means of communication and to the overwhelming challenge to come to terms with an alien environment.

Prolonged immersion presents additional problems related to orientation: to the gravitational vertical axis, vision, hearing and the sensory perceptions in

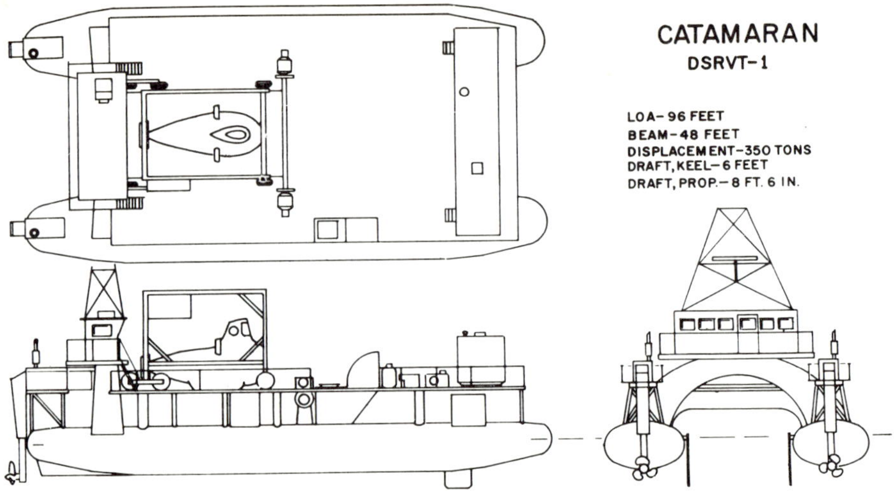

Fig. 16 WHOI catamaran outboard profile

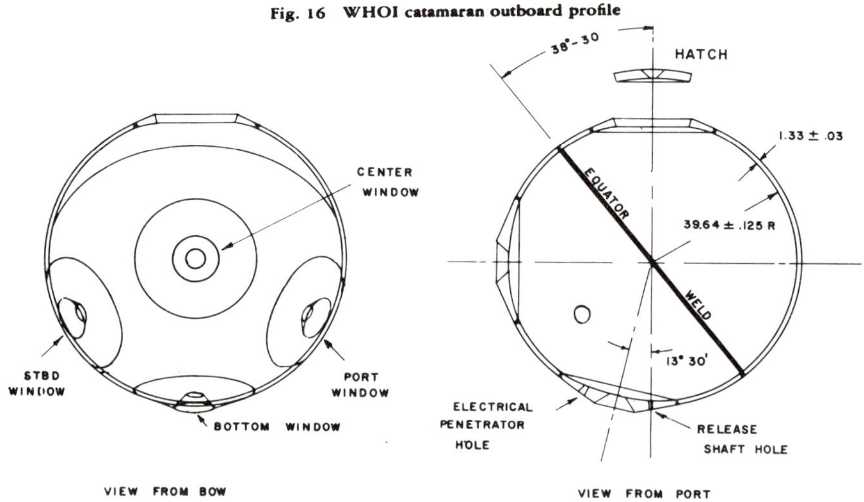

Above: Surface support system for Alvin.

Below: Personnel sphere for Alvin.

general. Reorientation requires the provision of critical stimuli by artificial means.

Perception of the gravitational vertical axis and the subsequent sense of position in space is provided by receptors which are divided into three groups: visual, tactile and static. The static receptors are the organs, located in the labyrinth of the inner ear, which are sensitive to gravity and comparable mechanical field forces. They consist of two separate organs, the semi-circular canals and the cochlea. In the absence of visual and tactile stimuli, however, persons with labyrinthine damage are easily disorientated. Although these organs operate under immersion, any form of sensory deprivation, temporary or permanent, is liable to alter their performance, and without optical or auditory stimulus, the

The Piccard-Grumman Mesoscaphe Ben Franklin (*PX-15*). *In June 1969, the PX-15 was towed to the Gulf Stream outside Palm Beach, Florida, where it submerged and began a silent drift with the Gulf Stream up to a point off Cape Cod, Newfoundland. During the six-week journey, the crew of six (three pilots and three scientists) were able to observe ocean life on film and by making acoustic recordings from special hydrophones.* (See also *pages 72 and 73.*)

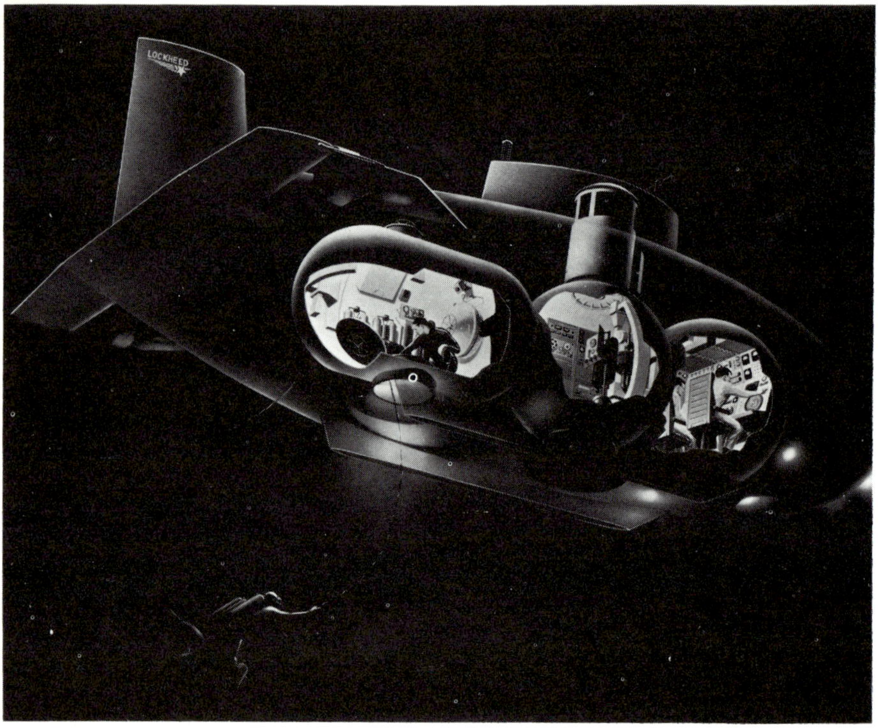

Above and Right:
Lockheed submersible Deep Quest *was launched in June 1967. The crew of the* Deep Quest *planted the American flag on the bottom of the Pacific ocean on 28 February 1968, as the submersible reached the record depth of 8,310 ft. Only observation bathyscaphes with limited manoeuverability and working capacity have penetrated deeper. The pressure hull is made of titanium alloy and steel nearly an inch thick to withstand the water pressure which, at 8,000 ft, is approximately 3,500 psi, or 230 times atmospheric pressure. The hydrodynamic outer hull is free-flooding and made of aluminium. The electrical power is capable of 24 hours continuous submerged operations at speeds up to $4\frac{1}{2}$ knots. The life support system is equipped to last 48 hours primarily as a safety factor, but this large margin enables* Deep Quest *to be used for developing advanced underwater search and recovery operations. There is a facility for a hyperbaric chamber for divers who must be maintained at the ambient water pressure, whereas the crew who are not required to work in the water are freed from the problems of pressurisation in a separate self-contained chamber at normal atmospheric pressure.*

individual under immersion is unable to find 'the direction of surface'.

The density of water and the blood and other body fluids are closely comparable, and the effect of gravity on the body organisms under water is negated. Together with this, oceanauts are neutrally bouyant under water (with the use of lead counterweights), and are effectively weightless. Special training is required,

therefore, for movement and work under water. The spatial co-ordination of oceanauts must adjust to accept the new dimensions of space and movement. Oceanauts must also periodically return to their habitat and the influence of gravity; here the spatial co-ordination of the interior becomes particularly relevant, and will affect the orientation of the occupants. Disorientation arising from unsympathetically co-ordinated space inside the habitat may lead to serious psycho-physiological stresses in prolonged dives.

Lockheed submersible Deep Quest.

The technology used in the design and development of most submergence hardware is fairly conventional. Existing and well-proven methods are used as far as possible, since the research and development expenditure is considerably limited compared, for instance, with programmes involved with the exploration of space. It is sometimes suggested that the design of the latest submergence hardware is merely an extension of existing techniques in conventional naval architecture and submarine construction, but the experimental programmes to date have only involved conventional techniques because of economic factors concerned with carrying out the programmes efficiently. Quite often, valuable innovations and techniques have to be omitted from experiments because the reliability and cost of a more conventional (perhaps less satisfactory) method is more appropriate in the pioneering stages of undersea programmes.

However, many developments of space programme technology in materials, design method, and construction have been used very successfully in submergence hardware, especially in underwater habitats. Many aerospace companies in the United States of America have become actively involved in the research and development of submergence hardware, and their efforts have a valuable new approach, free of the conventions that restrict shipbuilding establishments.

The most important single factor to emerge from the 'space company' with involvement with underwater experiments is the introduction of composite materials

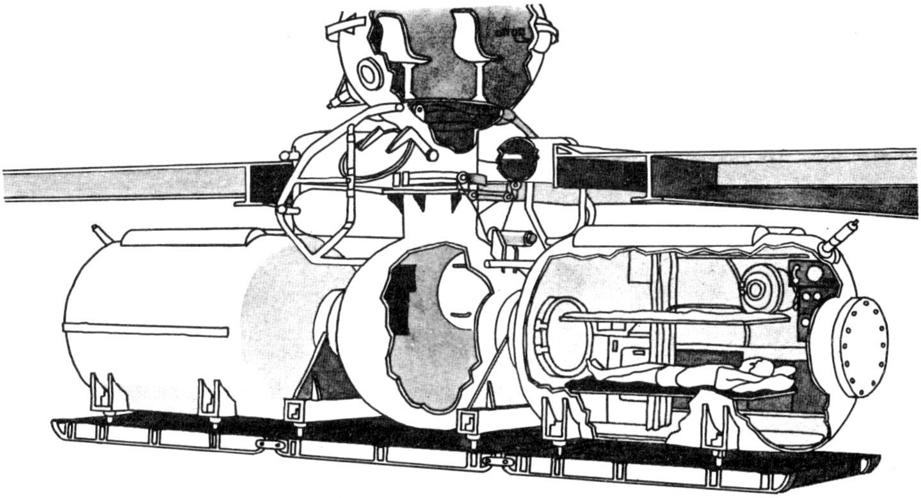

Above: A deck decompression system developed for the U.S. Navy.
Below: Sealab 3: *deck decompression chamber.*

into the construction of hardware, which, within the normal frame of marine construction, would be limited to steel or an alloy. The discovery that glass is extraordinarily resistent to forces of compression has led to its development for use in the construction of underwater habitats, in conjunction with panels of titanium alloy steels which provide the frame into which the glass panels are

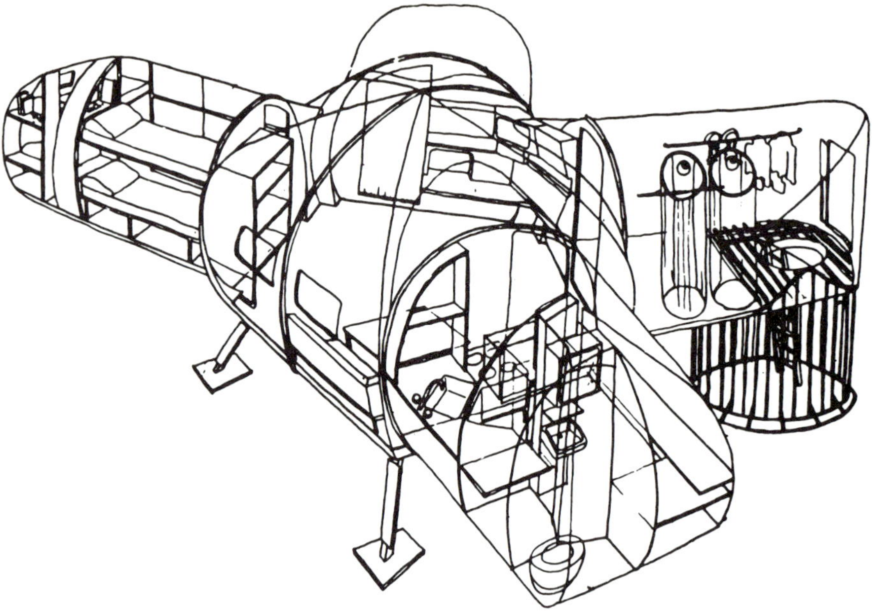

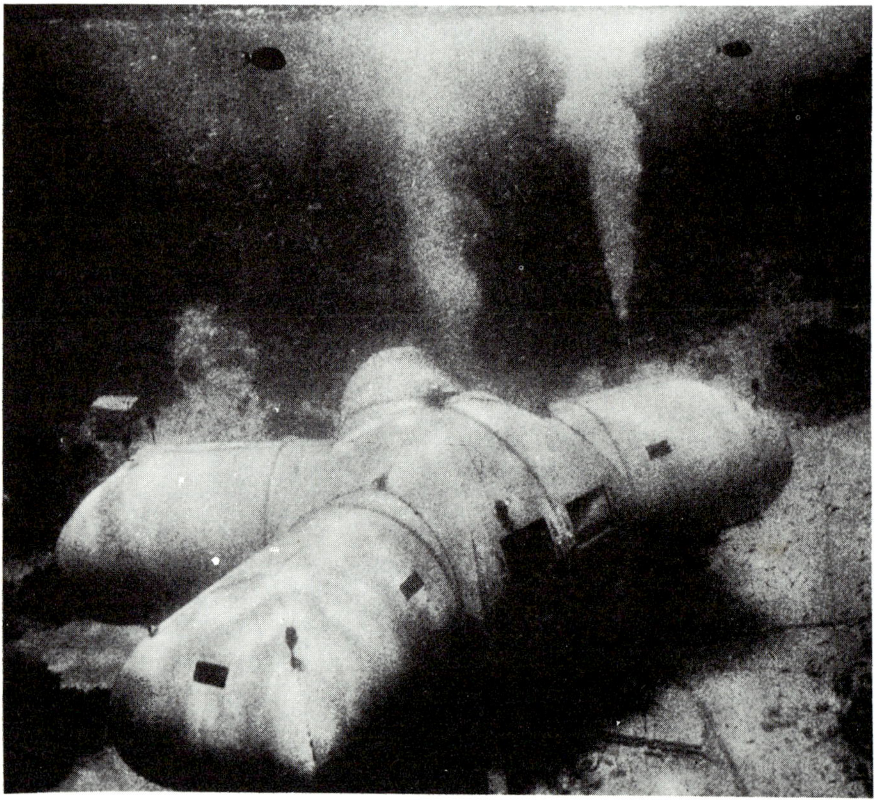

bound together. Many of the life-support systems used in submergence hardware were originally developed within the space programme of N.A.S.A. (National Aeronautics and Space Administration). This interrelationship of ocean and space technology, and to some extent the similarity of the psychological problems involved, has led to N.A.S.A. being specifically concerned with an experiment in living under water, which took place in 1969. It must be appreciated, though, that despite the use of materials similar to those of space technology, a common design approach, and to some extent the same construction techniques, many of the problems of ocean technology are quite unique, and some characteristics prevail for which no analogy can be drawn.

There are two different kinds of underwater systems from which most submergence hardware is developed. One involves the construction of units which are only subject to the pressure and general constraints of the ambient environment, and the other is quite specifically designed to withstand considerable changes in external pressure whilst maintaining a stable pressure inside. Ambient pressure systems have a great deal of flexibility both in the form they have to take and the materials from which they must be constructed, as the constraints

Left: Precontinent 2.
Below: Deep cabin divers on Precontinent 2.

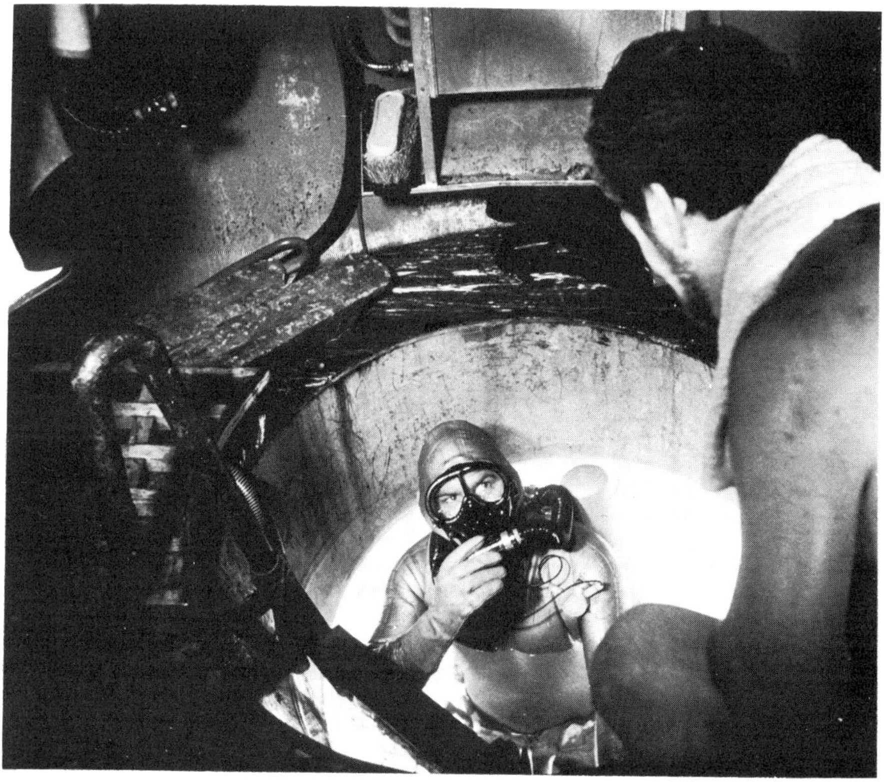

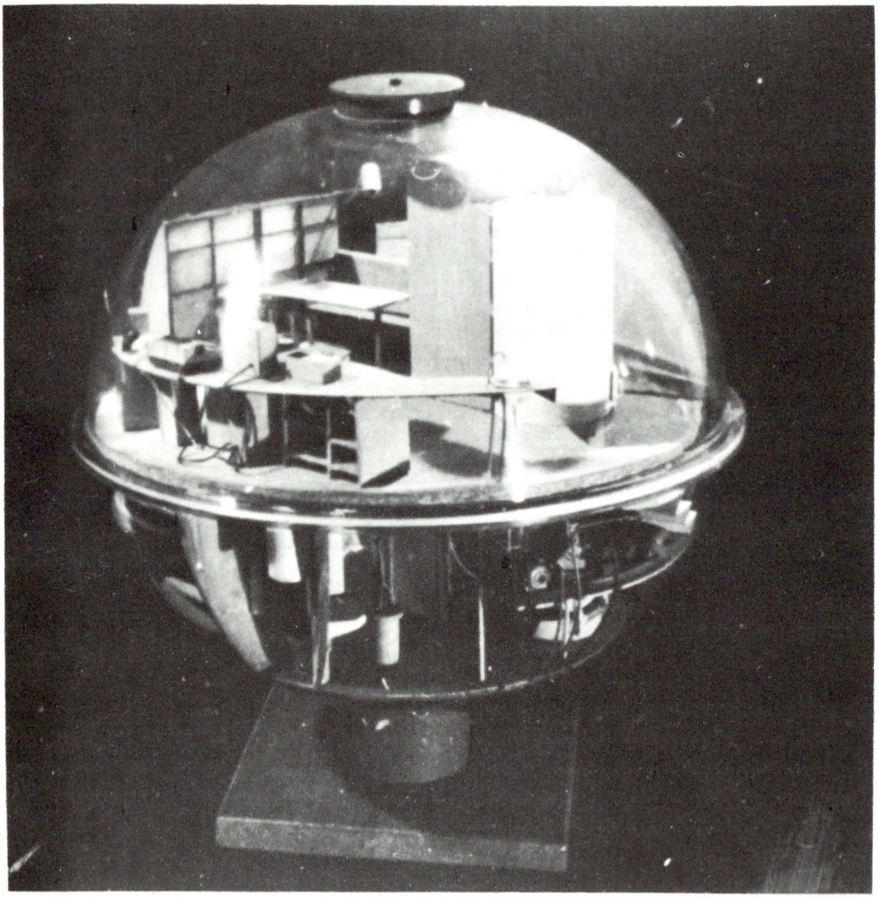

Above: Personnel transfer capsule used in the Sealab series of experiments.
Above left: Model of Precontinent 3.
Below left: Aquanauts eating on board Precontinent 3.

do not involve large changes in the pressure exerted on the structure. The pressure-resistent structures, however, are much more limited in the form they must take, partly because the ideal pressure-resistent shape is the sphere in which the stresses are equally dissipated from any point. Also, existing technology for pressure-resistent containers is limited to the use of spheres and cylinders. Both cylindrical and spherical forms have been used as the governing shape for a number of undersea experiments.

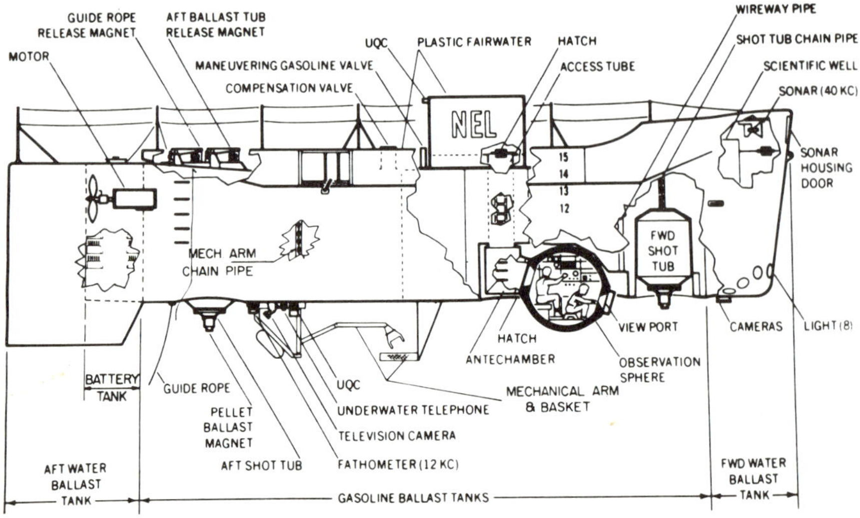

Above and right: Trieste.

Below: A diagram of Trieste 2.

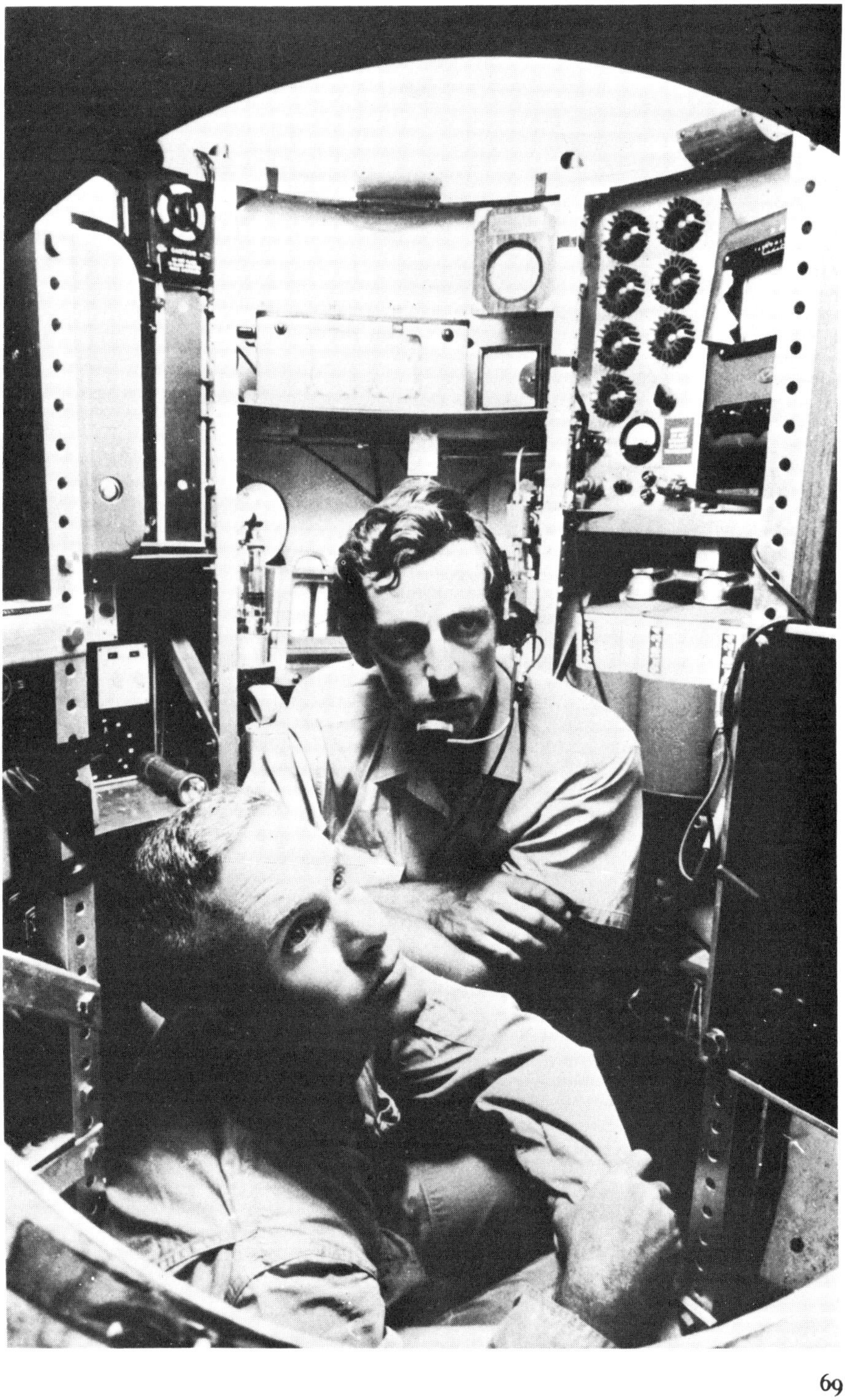

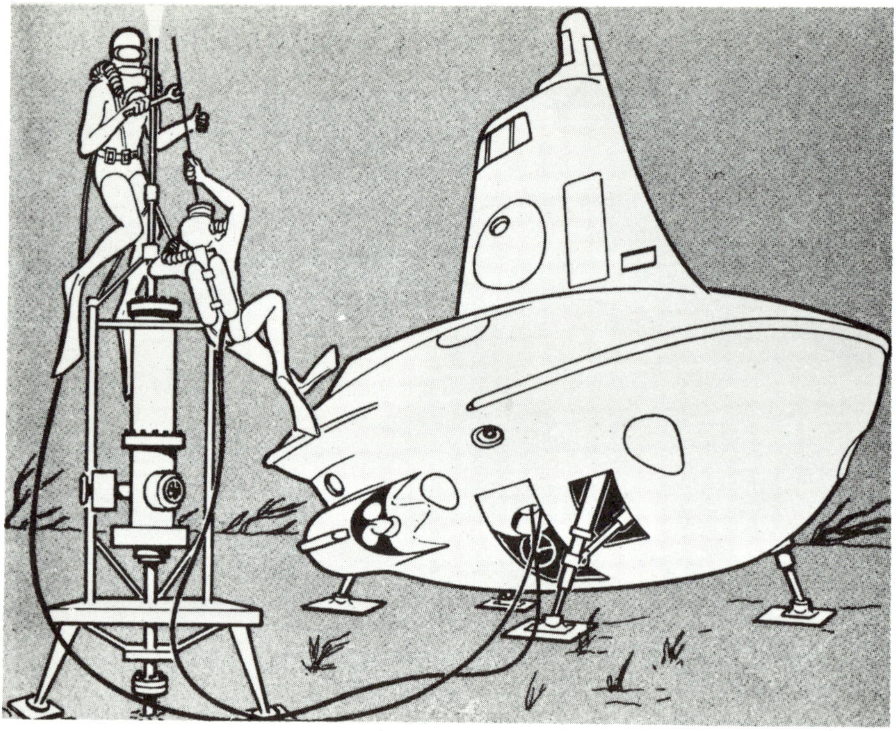

The submersible Argyronete *is being constructed by the Centre d'Etudes Marines Avancées in Marseilles, and should be completed in Autumn 1971. It will weigh 230 tons, have a depth capacity of around 600 m, with a crew of ten men, four of whom are divers in a separate hyperbaric chamber. The speed of the vessel submerged is estimated to be 4 knots and 6 knots at surface, its range operation about 400 miles. Duration underwater is limited to two days independent of an auxiliary power supply.*

The most impressive example of the freedom with which an ambient pressure system may be designed is the *Precontinent 2* experiment of Commdt. Cousteau (see Appendix 2), though this extraordinary form, 'Starfish House', was arrived at by completely functional criteria. From previous experiments and data, it had been determined that within the budget available the habitat should be constructed as simply as possible from the most reliable conventional techniques; the choice of welded steel was then inevitable. The forms that are most easily fabricated in steel are cylindrical and spherical, so the design utilised 'domes (hemisphere) and cylinders'. From an assessment of the general requirements of the experiment it was determined that four to five different areas were required; the largest of these became the dome, and the subsidiary area became the cylinders attached to it.

A number of inflatable dwelling units have been developed for use under water, where the pressure of the atmosphere inside is partially greater than the

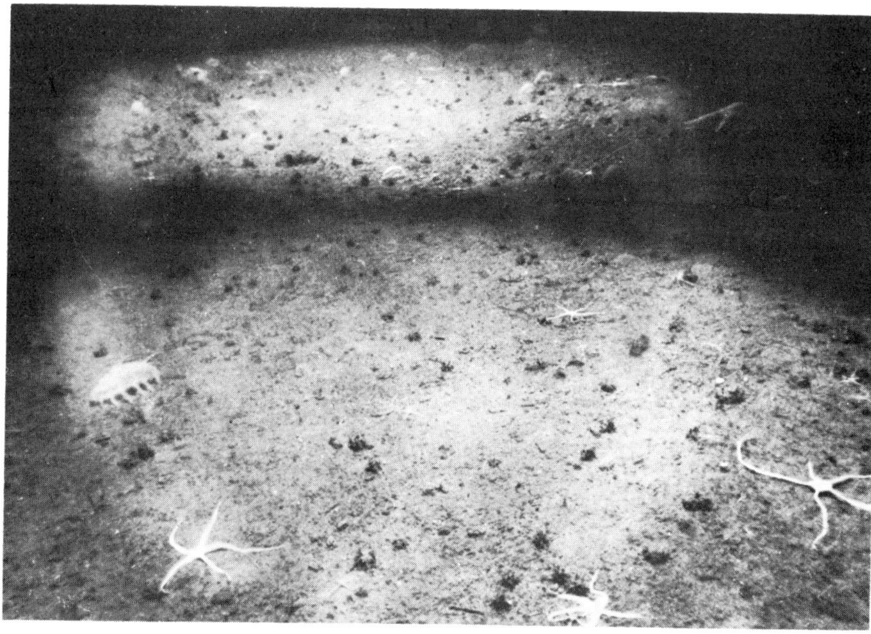

The ocean bottom at Challenger Deep.

ambient pressure in order to keep the walls open. The problem with this kind of system is that as yet no suitable material has been developed, since most rubberised materials are permeated by gases under only slight variations in pressure, and much of the gas mixture would be lost into the water through the skin of the habitat. This loss would be tolerable at shallow depths when using air, but at greater depths, where gas mixtures involving large proportions of expensive inert gases like helium are being used, this kind of loss through the skin of the habitat would be economically disastrous.

Pressure-resistent systems have been developed using both spherical and cylindrical forms. The difference from ambient-pressure habitats is that here the form is quite definitely determined at the start, and the life-support and experimental systems have to be incorporated within it. As far as is possible, all equipment which could prove dangerous is placed outside and isolated from the occupants of the habitat. The layout of the equipment and systems within the craft is then governed by the importance of each item, and where it will fit within the pressure hull. Experimentally, the approaches to this problem differ radically. Sealab has developed a complex and sophisticated system based on a cylinder into which many advanced systems have been fitted, quite regardless of whether they create a visually congenial environment for the aquanauts. The Cousteau unit, however, seems to have accepted that an organised interior has an important psychological advantage over one that appears chaotic and incomprehensible, even if it is not so. The spherical habitat of *Precontinent 3* (see Appendix 2) is divided into two sections; the upper hemisphere contains the living quarters

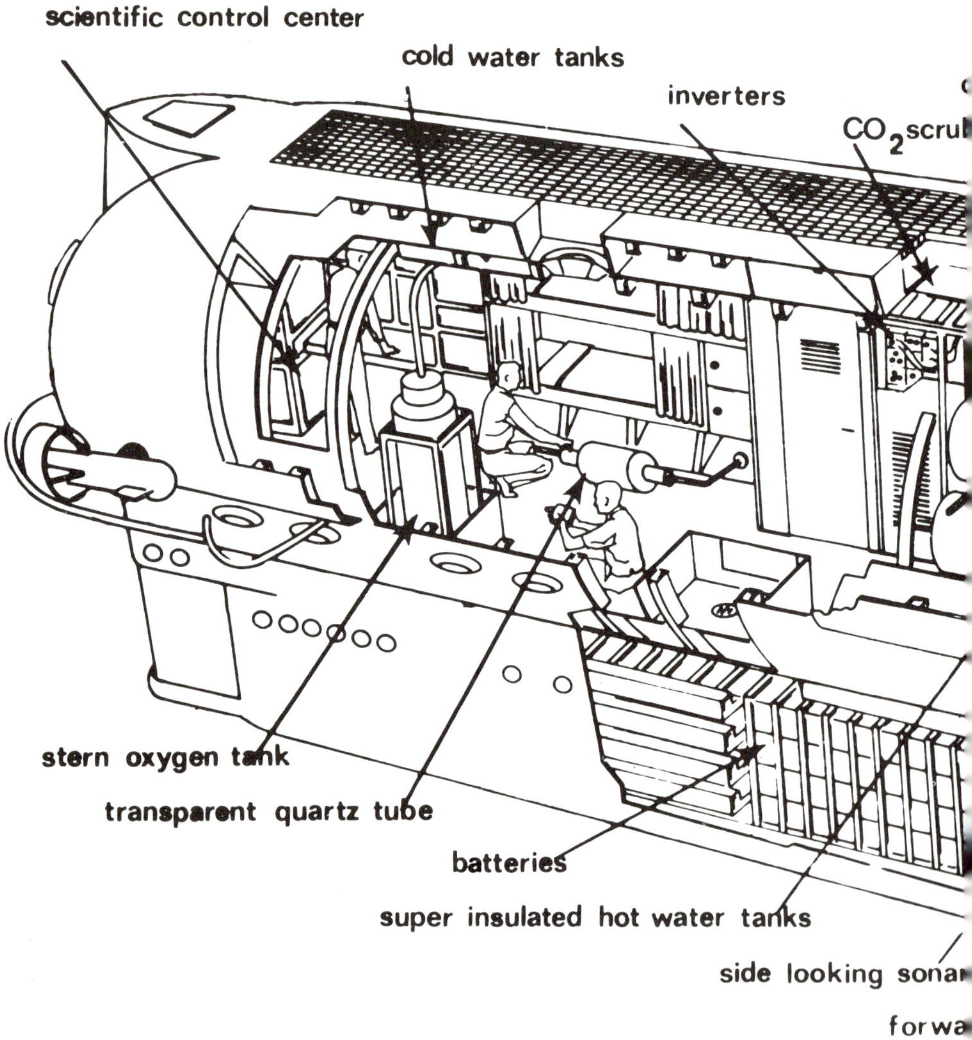

The Piccard-Grumman Mesoscaphe Ben Franklin (*PX-15*). (*see page 59*)

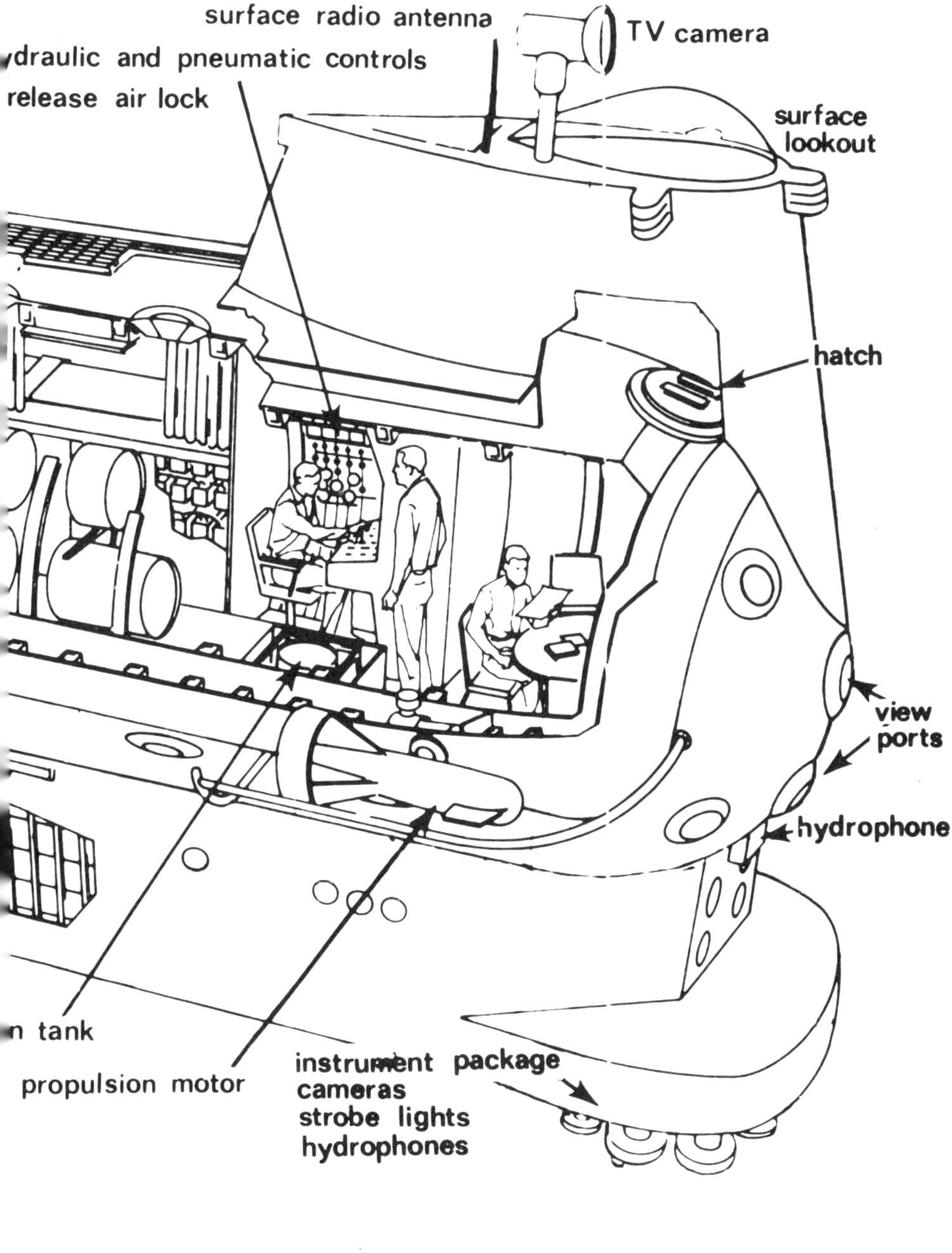
surface radio antenna
TV camera
ydraulic and pneumatic controls
release air lock
surface lookout
hatch
view ports
hydrophone
n tank
propulsion motor
instrument package
cameras
strobe lights
hydrophones

and control room; the lower is the operations area from which the underwater work takes place.

The latest submersible craft have incorporated into their pressure-resistent system a provision to carry divers in a separate compartment maintained at the pressure at which they are required to work under water. This kind of craft gives a considerable degree of manœuvrability, as well as being able to carry specialist non-diving personnel at normal atmospheric pressure in the main part. Undoubtedly from this kind of vehicle the working underwater craft of the future will be developed; the need for stations and bases fixed on the bottom will come later when the exploitable resources have been discovered and assessed.

The value of operational hardware designed with some sympathy for the occupant does not concern the experimental stages of underwater programmes, because volunteer oceanauts are prepared to endure virtually anything, however inconvenient, to see the experiment through successfully. This intensity of motivation cannot be expected from individuals making underwater work their career. No longer pioneers, they will merely be carrying out tasks that will have become relatively mundane and routine. It is crucial that this category of worker in the future does not suffer from the lack of a satisfactory working environment, and is not subjected to the demands of commercial exploitation.

Below: The bottom at the deepest part of the ocean.
Right: Observing marine life (from 20,000 Leagues under the Sea *by Jules Verne).*

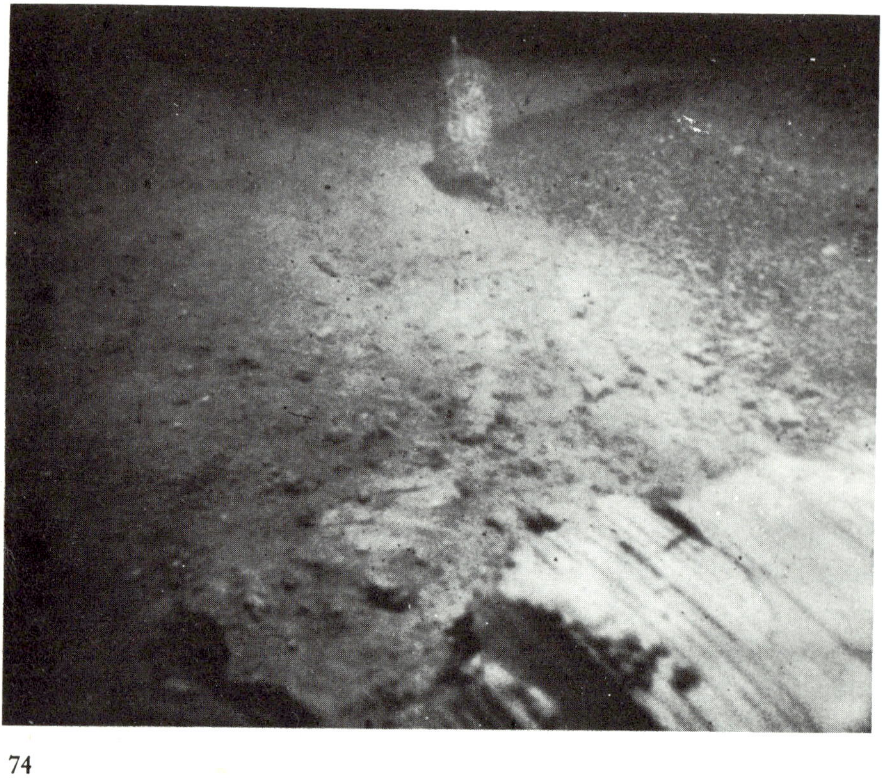

Within the solar system the earth is the only planet which has oceans. Jupiter and the other large planets have small solid cores, covered by gases which have become liquefied under pressure, and have relatively enormous atmospheres. The inner planets which are closer in size to the earth have only very superficial similarities to it.

The repetitive pattern of erosion and sedimentation suggests that the land masses and oceans have always been present on the earth. What is perhaps more surprising is the constancy of the total volume of water throughout the earth's evolution. The sea level has of course varied; during the ice ages, when a great quantity of water was locked up in the ice sheets on the continents, the sea level was much lower than it is now, and the continental shelves of Europe and North America were 'surfaced'. Often the oceans have advanced over the coastal plains, but they have never covered all the land or even most of it. The mechanism of this balance is as yet unknown. It is quite conceivable that water would be gradually expelled from the interior of the earth and that the oceans would grow steadily larger; conversely that water would be dissociated into hydrogen and oxygen in the upper atmosphere, and that the hydrogen would escape, leading eventually to the drying up of the oceans. These processes either do not occur, or are in equilibrium and cancel each other out.

The problem is complicated by the apparent scarcity of neon in the atmosphere; in the sun the occurrence of neon is comparable with that of oxygen. The earth's neon was presumably lost during the creation of the planets, because it does

Hypsographic curve showing the area of the earth's solid surface above any given level of elevation or depth. The solid black is the frequency distribution of elevations and depths at 1,000 m intervals.

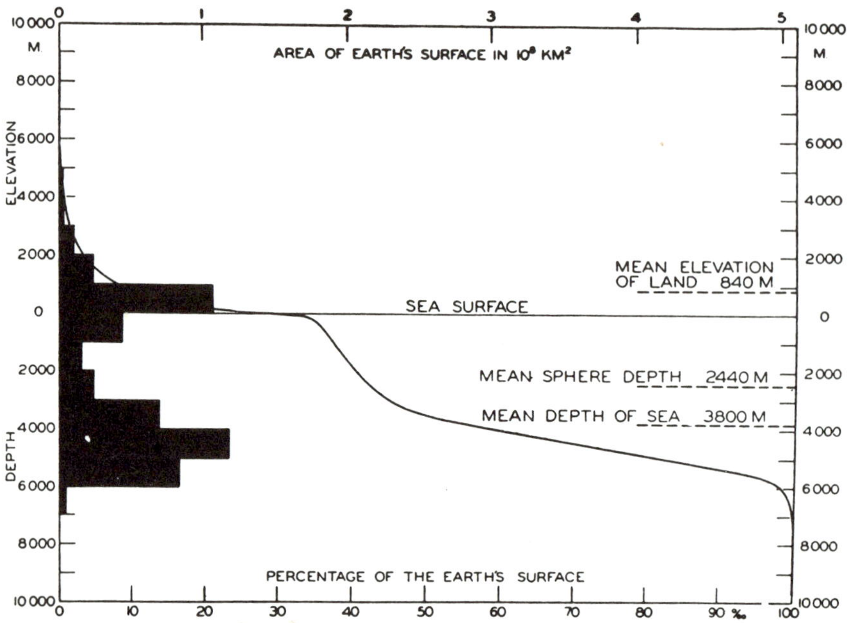

not easily form compounds. Perhaps water would also have escaped (since it is of lighter molecular weight than neon), but for the fact that it readily makes compounds. Before the earth had formed, water was probably combined in hydrated silicates and released as a liquid from these compounds during the first billion years of the planet's history.

As the evidence of the earth's history is almost non-existent these speculations are quite obviously vulnerable, but the study of the structure of the ocean floor recently – particularly during the International Geophysical Year (1957-8) – has contributed largely to the understanding of the origin of the oceans. The realisation that the structure of the ocean floor is quite different from the continental land mass was a crucial step in this direction. The 'undersea mountain ranges' that were discovered are not comparable with their land counterparts. The undersea mountain range is a vast system of mid-oceanic ridges all round the world. It is composed not of folded sediments as the land ranges are, but entirely of igneous rocks of basalts which have emerged from the interior of the earth. The undersea mountains have a covering of sedimentation in many places, which apparently has been deposited rather than folded or compressed.

A crack-like valley runs along the crest of the mid-oceanic ridges for most of

Distribution of land and ocean over the earth's surface.

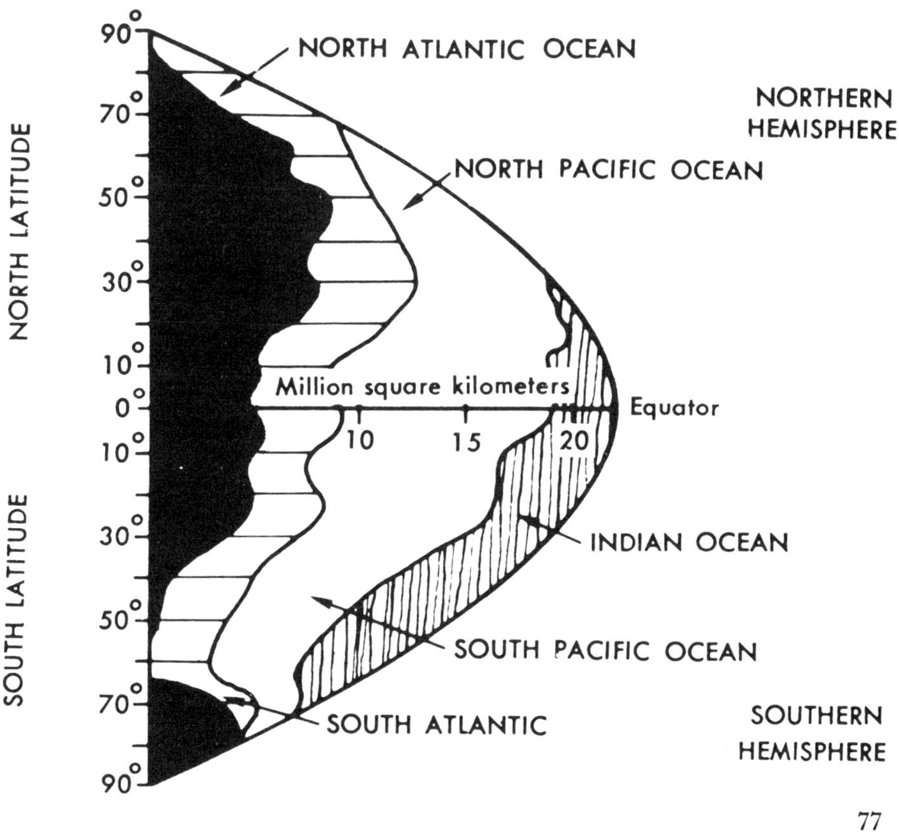

The world's oceans: of the total mass of water present on the planet only 13.5 per cent may be directly attributed to continental ice, lakes, rivers, water vapour in the atmosphere, and in rocks and sediments; the remaining 86.5 per cent of mass of water is contained in the world's oceans which cover almost 71 per cent of the earth's surface at an average depth of 3,730 m.

The oceans of the planet are in fact one ocean, but are divided along degrees of latitude into the Pacific, Atlantic, and Indian for reasons of academic convenience. Of the marginal seas which consist of the Arctic, Carribean-Gulf of Mexico, Mediterranean, Black Sea and the Caspian Sea, only the Caspian is completely landlocked, but is thought to have been once connected to the world ocean, being cut off by changes in the sea level and land movements.

The projection illustrates the extent of the Continental Shelf (unshaded area) which is adjacent to the continental land mass (in black). The ocean floor is shown in the horizontal-lined area, the mid-oceanic ridges in the diagonal-lined area. The deep ocean trenches are indicated by heavy black lines and can be identified as follows: 1) Peru-Chile Trench; 2) Aleutian Trench; 3) Kuril-Kamchatka Trench; 4)

Japanese Trench; 5) Philippine-Marianas Trench; 6) Kermadec-Tonga Trench; 7) Java Trench; 8) Puertu Rico Trench.

their length, and it is here that the new ocean floor is being formed today. From the study of numerous earthquakes along this crack it is apparent that the two sides of the ridge are moving apart and that the crack would steadily widen were it not being filled with material from within the earth's crust. As the rocks on the two sides move away and the new rock solidifies in the crack it becomes magnetised in the direction of the earth's magnetic field. For at least the past 10,000 years, and probably longer, the north magnetic pole has been close to its present position, so that the magnetic field is to the north and downward in the northern hemisphere, and south and upward for the southern hemisphere. As the cracking and spreading of the ocean floor goes on, a strip of magnetised rock is produced. In the course of several thousand years the earth's magnetic field reverses and consequently the next effusion of lava becomes magnetised in the reverse direction, so that a strip of reversely magnetised rocks is built up between the two split halves of the previous strip. These reversals succeed one another at widely varying intervals; sometimes the change comes after 50,000 years; often there is no change for a million years and occasionally there is no reversal for 20 million years, as in the Permian period. The process of ocean floor spreading is recorded in the oceans of the world by a sequence of reversals of magnetic field in the rocks composing the ocean floor. These reversals of field can be measured by oceanographic ships equipped with magnetometers, which are towed behind the ship to determine the magnetic polarisation of the rock on the ocean floor.

The speed of spreading on each side of a mid-oceanic ridge varies from less than a centimeter a year to as much as 8 cm a year. The fastest rate of spreading is from the East Pacific Rise and the slowest rates are from the Mid-Atlantic ridge. The rate at which new terrestrial crust is produced at the crack-like valley of a ridge is the sum of the rates of spreading on its two sides. This rate, often the same for both halves of the split, may be as much as 16 cm a year, which, geologically, is fast. At 16 cm a year, the entire floor of the Pacific Ocean, which is about 15,000 km wide, could be produced in 10 million years.

The mid-oceanic ridges are not continuous, but are cut into a number of sections by fracture zones. A study of the earthquakes on these fracture zones shows that the separate pieces of ridge crust on the two sides of the fracture zone are not moving apart. They remain fixed with respect to each other, while on each side the crust moves away as a rigid body. Fractures of this kind are called transform faults.

The regularity of the magnetic pattern suggests that the ocean floor can move as a rigid plate over areas several thousand kilometres across. The thickness of these rigid moving plates is uncertain, but could be nearly 70 to 100 km. As such the greater part of the plate will be made of the same material as the upper part of the earth's mantle – probably of peridotite, a rock which is largely composed of olivine, a silicate of magnesium and iron. The basaltic rocks of the oceanic crust will form the upper 5 km or so of the plate, with a thin layer of sediment on top.

The continents can be regarded as part of the same plate as the adjacent ocean floor; the rocks of the continental crust ride on top of the plate and move with it. At some places though, like the Pacific coast of South America, the oceanic plate dives under the continent at an angle of about 45°. On the upper surface of this sloping plate, numerous earthquakes occur, quite shallow ones near the coast and inland under the continent, some others as deep as 700 km.

The formation of ocean deeps along the edges of continents where a plate turns down is not precisely understood, but if a plate goes down it follows that a depression is left in the ocean floor. As the plate goes down some of the sediment on its surface is probably scraped off and piled up on the land side of the ocean deep. This sediment may later become part of the mountain ranges that usually appear adjacent to ocean deep trenches on the edges of the continents. The mountain ranges bordering the continent commonly have volcanoes, as have the Andes. The lavæ from the volcanoes are frequently different from the lavæ of the mid-oceanic ridges because they contain larger amounts of silica.

Walking on the bottom (*from* 20,000 Leagues under the Sea *by Jules Verne*).

The overall pattern shows that the majority of the earth's surface is divided into six plates which move as rigid bodies, the new material for them being produced from the upper mantle by lava emerging from the crack-like valleys along the crest of the mid-oceanic ridges. Plates are destroyed at oceanic trenches by plunging deep into the mantle where they become mixed once again with the material from which they originated.

That the earth's surface developed by the formation of plates, their movement, and destruction, is not verifiable, but can be broadly tested by drilling. The work of a group of American universities has shown that the rate at which the ocean floor is spreading has been constant for approximately 70 million years. The process of consumption of oceanic crust at the edge of a continent may proceed for tens of millions of years, but if the plate that is being consumed carries a continental fragment, then the consumption must stop when that fragment reaches the trench and collides with the continent beyond it. Because the fragment consists of relatively light rock, it cannot be forced under a continent, but either becomes welded to it, or forces it to break up.

The theory that all the continents were collected in a single unit for millions of years and were then broken apart and set moving during the last 100 million years seems unlikely. It is much more probable that the processes apparent today were always in action, and that throughout geologic time there have been moving plates carrying the continents, which must have split many times, formed new oceans and sometimes must have collided.

The oceans and atmosphere are intrinsically linked. The atmosphere drives the circulation of the oceans and influences the composition of seawater; to a great extent the atmosphere in turn owes its nature to and derives its energy from the ocean. Indeed, very few phenomena of physical oceanography are not considerably influenced by the atmosphere, while there are few atmospheric phenomena for which the ocean plays an unïmportant part.

The ocean and atmosphere are the two unique thin fluid films which cover the earth, in which all life is sustained and whose nature and movement completely determine the conditions of the environment. Although the ocean's effect upon the temperature of the atmosphere is the more apparent interaction, the ocean is in turn cooled when it gives up heat to the atmosphere. The density of the ocean's water is controlled by two factors: temperature and salinity. Evaporative cooling tends to make the water denser: it lowers the temperature and, since evaporation removes water but comparatively little salt, it also increases the salinity. If the surface water becomes denser than the water underlying it, vertical convective mixing takes place. In a few places in the ocean, this can be so vigorous that the water will sink very deep, occasionally right to the ocean floor. Once cold water has dropped so far, it is heated from above so slowly that it remains at the same depth for a very long time, with little change in temperature. This heavy cold water constitutes the bulk of the ocean's water.

The atmosphere drives the circulation of the oceans and the heat supplied from the ocean is largely responsible for the release of energy for the atmosphere. Obviously there is a considerable amount of feedback between the two systems. The atmospheric patterns determine the flow of the ocean, which in turn affect where and how much heat is released to the atmosphere. The atmosphere-flow systems determine how much cloud cover there will be over certain parts of the ocean and therefore how much, and where, the ocean will be heated by the sun.

This system, unfortunately, is not especially stable. Each part of the world has its own abnormally cold or mild winters and dry or wet summers. The persistence of such anomalies over many months almost certainly involves the ocean, since the characteristic time constraints of atmospheric phenomena are much shorter. Long-term climatic variations, such as the great Pleistocene glaciations, are likely to have involved changes in the oceans' circulation.

The ocean-atmosphere system has a number of near stable configurations. There may be a number of different patterns in which the atmosphere can drive the ocean so that heat is released in the right places and quantity, allowing the pattern to continue. The atmosphere is extremely turbulent, so that its equilibrium is constantly being upset, but since there is stability in the system, forces come into action to restore the condition after each disturbance. As there are a number of different stable configurations, it is possible that a particularly large disturbance might take the system from one stable configuration to another, either more or less favourable than the present one.

The resources of the world's oceans include seawater itself, which acts as a processing plant to convert solar energy into protein; it is also a large store of dissolved minerals and fresh water, serves as a waste disposal system, and

Section of a photograph from N.A.S.A.'s ATS III earth synchronous satellite, reproduced in black and white (photograph taken 18 November 1967, 1567 GMT).

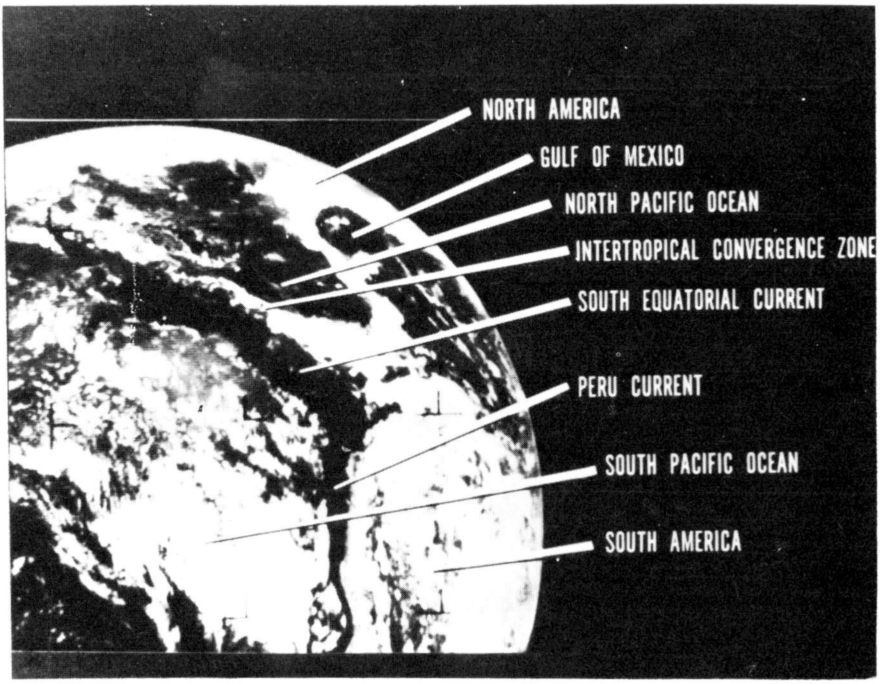

provides a source of energy from the action of the tides. There are deposits on the ocean floor of minerals, sediments and rocks, together with quantities of fossil fuels (oil, coal and gas), and minerals in the floor itself. The coasts are a potential resource that is only too vulnerable to reckless exploitation and irrevocable damage by man. Because the oceans make up the greater part of the planet, any statistical information on their resources and potential resources is invariably impressive. It is important to understand, however, that the significance of these resources and their long-term relevance to man will necessarily involve exploration and development that is fundamentally different to the standard techniques and systems used today. What is perhaps most important is that resources of the oceans are the common property of the inhabitants of the planet, and although most will inevitably be privately developed, there is just a glimmer of hope that some kind of international co-operation might take place to enable these resources to be used for the benefit of mankind in general.

The seawater in the oceans, which constitute 330 thousand million cubic miles of water, make up the largest single mineral resource in the world. The solids dissolved in the seawater are about 35,000 parts per million, so that in every cubic mile of ocean there are about 165 million tons of solids.

Although most of the chemical elements have been detected in seawater (it is quite likely that all chemicals are present in the oceans), only common salt, magnesium and bromine are extracted in significant quantities at the moment.

The elements present in seawater.

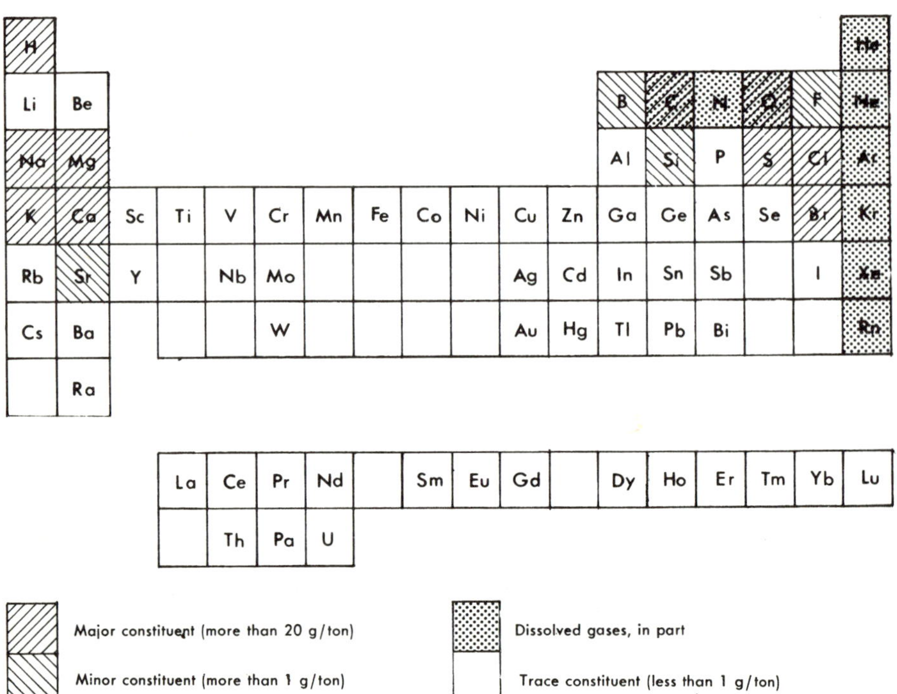

Magnesium is the third most abundant element in the oceans, and the most valuable of those currently extracted. The ocean is the source of 70 per cent of the world's production of bromine, despite the fact that its concentrations are as small as 65 parts per million.

The economic recovery of other chemicals from seawater is not possible with contemporary technology. The value of the 17 critical metals present in a cubic mile of seawater is significantly less than the cost of operating any existing system to extract them.

The extraction of fresh water from seawater is also a process which has been difficult to carry out economically, although the demand for fresh water is increasing rapidly enough to make the development of desalinisation systems a worthwhile proposition. It is important, however, to recognise that all the resources in seawater are not open to indiscriminate extraction, as they make up a composition which is important to the ecological function of the oceans, and any significant alteration in the composition of seawater would have catastrophic effects on marine life.

There was a time when the oceans could be safely utilised as a waste disposal area, where sewage and effluents would be rapidly dispersed, broken down and diluted. Regrettably this is no longer possible with a real degree of safety. The concentration and the bulk of material disposed, together with the new kinds of waste products that cannot be broken down, or are actual contaminants, is a serious threat to the coastal areas which are anyway subject to great demands from human activity.

Apart from polluting coastal areas, concentrations of contaminants are rapidly forming in the open ocean. These are less easily isolated and often have far more serious consequences. The lead from petrol additives introduced 45 years ago has increased the lead concentrations in the Pacific by ten times. Toxic chemicals and insecticides like DDT residues have appeared many thousands of miles from where they were deposited, having drifted with currents and winds. Radio-active fallout from atmospheric nuclear test explosions can now be isolated in any part of the oceans of the planet. It is tragic that this sort of myopic behaviour continues, and, though the dangers are periodically recognised, the crucial issues are ineptly avoided again and again.

The mineral resources of the ocean floor are highly localised and scattered deposits; they include fluids and soluble minerals such as gas, oil, sulphur and potash that can be extracted from the surface through boreholes. Subsurface deposits, such as the metals which are found in veins, have up to now only been mined from tunnels that originate on the land. Some surface deposits can be dredged, such as the heavy metals, glauconite (oyster shell and similar mineral compound), diamonds and authigenic minerals, nodules of manganese and phosphorite that have been formed by slow precipitation from the seawater. Of all these resources, oil and gas are by far the most valuable, and are likely to receive the greatest attention from future developers.

Sulphur, one of the world's major industrial chemicals, is found in the cap rock of salt domes buried within the sea-floor sediments. The sulphur is recovered quite simply and inexpensively by melting it with superheated water piped down from the surface and then forcing the sulphur liquid up with compressed air.

The only minerals on the ocean floor that are recognised to be of potential economic importance are manganese nodules. These are formed by the precipitation of the composing elements and seawater. The nodules are widely distributed, although the highest concentrations are in the Pacific Ocean. The composition of the nodules is approximately 24 per cent manganese; 14 per cent iron; 1 per cent nickel; ·5 per cent cobalt; and ·5 per cent copper. The combination of these minerals in the nodules makes them difficult to exploit commercially until an economic metallurgical solution to separate the metals is found.

Three undersea fountains of hot high density brines have been found in the Red Sea. These brines contain minerals in very high concentrations, and have almost 10 times as much solid matter as is normally present in seawater. Also they occur over sediments rich in heavy metals like zinc, copper, lead, silver and gold. Similar deposits may be characteristic of other enclosed basins associated, as the Red Sea is, with rift valleys.

Discoveries of new mineral deposits and potential resources are becoming increasingly frequent; chromite, zirconium, titanium and other heavy metals have been detected in various parts of the ocean floor. New oilfields are being discovered on the continental shelves, many of them of considerable economic value. It is also quite possible that the continental rises, which lie at depth ranges of 5000–18,000 feet and contain a much larger total volume of sediments than the continental shelves, may have significant oil reserves as well.

Man is now in the unique position of being able to calculate the ocean's resources, develop the technologies for successful exploitation, and guard against the reckless destruction of a colossal potential by short-term policies, and ignorant and inept politics. With luck, it may even be possible to correct the mistakes that were made long ago, and some of those that are being made today.

The most valuable aspect of the ocean is as a renewable resource of protein-rich foods and food supplements, and pharmaceutical chemicals. Full advantage should be taken of the biological processes of the oceans, with the knowledge that the system is a finely balanced one and that to upset its function substantially or in part would only compound further the problems of successfully harnessing the oceans of the world to meet man's growing needs.

The harvest from the oceans today is around 55 million metric tons per year. More than 90 per cent of this harvest is fish; the rest consists of whales, crustaceans, molluscs and some other invertebrates. Three quarters of the total catch is taken by as few as 14 countries, though nearly all coastal countries make significant catches. Fish is one of the few major foodstuffs showing an increase which has been accompanied by a changing pattern of use: less fish is being used directly as human food, and much more is being incorporated into agricultural processes, into animal food or fertilisers, to increase the potential yield from the land masses.

Knowledge of the living resources of the oceans is still far from complete. It is possible, however, to obtain a basic overall picture by comparing various available factors and making speculative deductions from them. There are

Right: Fishing with a green laser.

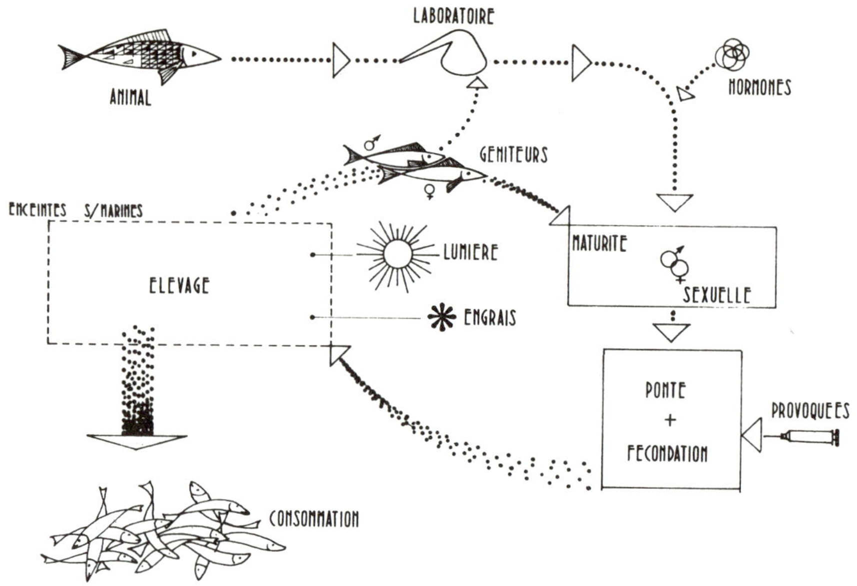

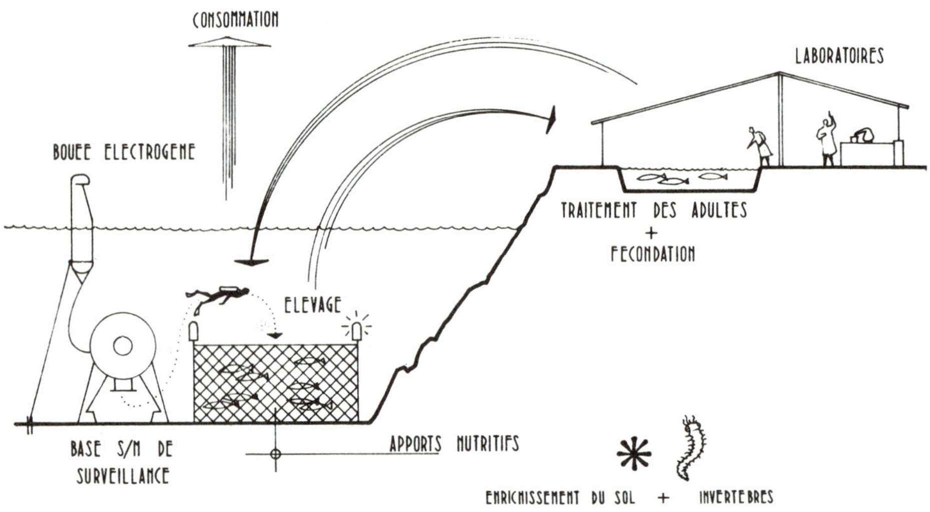

estimates of the food supply available to fish in a particular area, or of the primary production there, and from what is already known of metabolic and ecological efficiency actual fish production can be calculated.

According to available data these methods are being applied to the major groups of food fishes and it appears that the potential catch could be quite easily four times larger than the present one. Fishing operations, however, depend on locating areas of concentration of fish, and on accurate and efficient

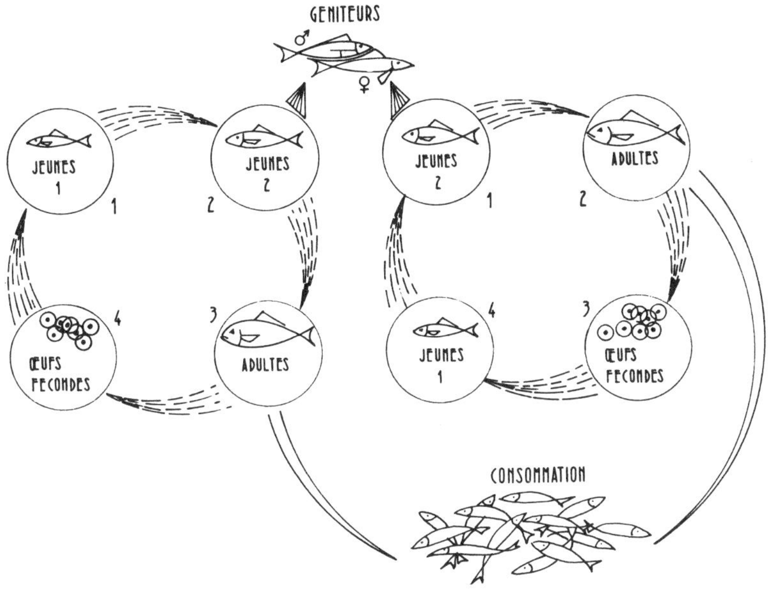

A hypothetical system for an undersea farm.

prediction of changes in these areas. The large stocks of open-sea fishes are productive in areas of divergencies, where water is rising from deeper levels towards the surface and where the surface waters are consequently flowing outward. Areas like this occur by the 'upwellings' off the western coasts of continental masses, where the seasonal winds and continental configurations combine to cause the surface layers of the ocean to be periodically enriched.

Divergencies are also associated with certain current systems in the open sea. Biological production is high in these areas because the nutrient salts needed for plant growth, and which are usually in limited supply, are renewed in the surface layers of the water. The blooming of phytoplankton, which is detectable in infra-red photographs of the oceans' surface, is associated rather with the fact that the water coming to the surface is cooler, than with its wealth of nutrients. A cool-water area of the ocean is characterised by seasonal peaks of primary production; the phytoplankton blooms are followed, after a time lag, by a large amount of herbivorous zooplankton that provides concentrations of food for large schools of fish. The size of this herbivore crop also varies according to the success of synchronisation with the primary production cycle. The advantages of being able successfully to predict variations in the conditions of the oceans would be considerable.

Recently the advances being made in the uses of earth-orbiting satellites for recording resources have led to the possibility of using a satellite to identify variations in ocean conditions pertinent to the production of food. Two satellite sensors, one operating in the infra-red wavelength, the other photographic,

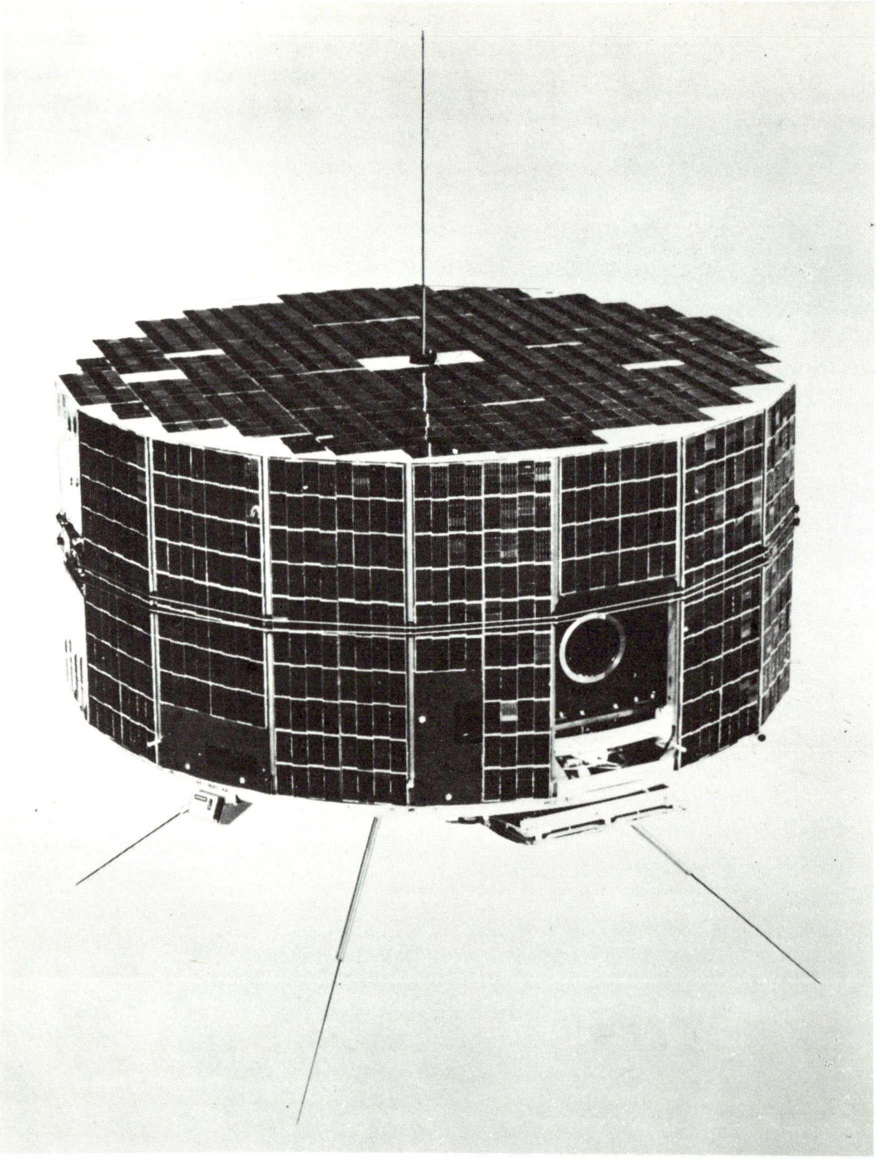

ESSA 6: automatic picture transmission satellite.
Right: The satellite Nimbus.

have shown well-known oceanographic features such as parts of the Gulf Stream. Earth synchronous orbit satellites, such as N.A.S.A's Applications Technology Satellite (ATS 2), are even more significant in that they permit a 'real time' synoptic analysis of ocean conditions. Providing that the data available to such

satellite systems can be usefully related to the factors controlling the production of food in the ocean, such systems could be developed for use on a world wide basis.

Contemporary knowledge of world ocean processes relies largely at the moment on data gathered by the individual ships and stations over the ocean's surface. To collect this data often enough to provide an accurate and unified synoptic analysis, and to co-ordinate it, proves extremely difficult. Although photographic and spectral scanning by satellite could provide equivalent or comparable data, surface stations are still required to gather data for specific areas.

The thermal structure of the oceans is fundamentally related to all marine processes, one of which is the migration of marine life. The dynamic nature of many of these large-scale thermal features means that they can only be effectively studied with frequent and repetitive measurements over a wide surface area. Breaks in sea surface temperature have been detected by aircraft in the visible infra-red and micro-wave regions of the electromagnetic spectrum. Infra-red cameras in aircraft have successfully recorded areas of strong thermal contrast

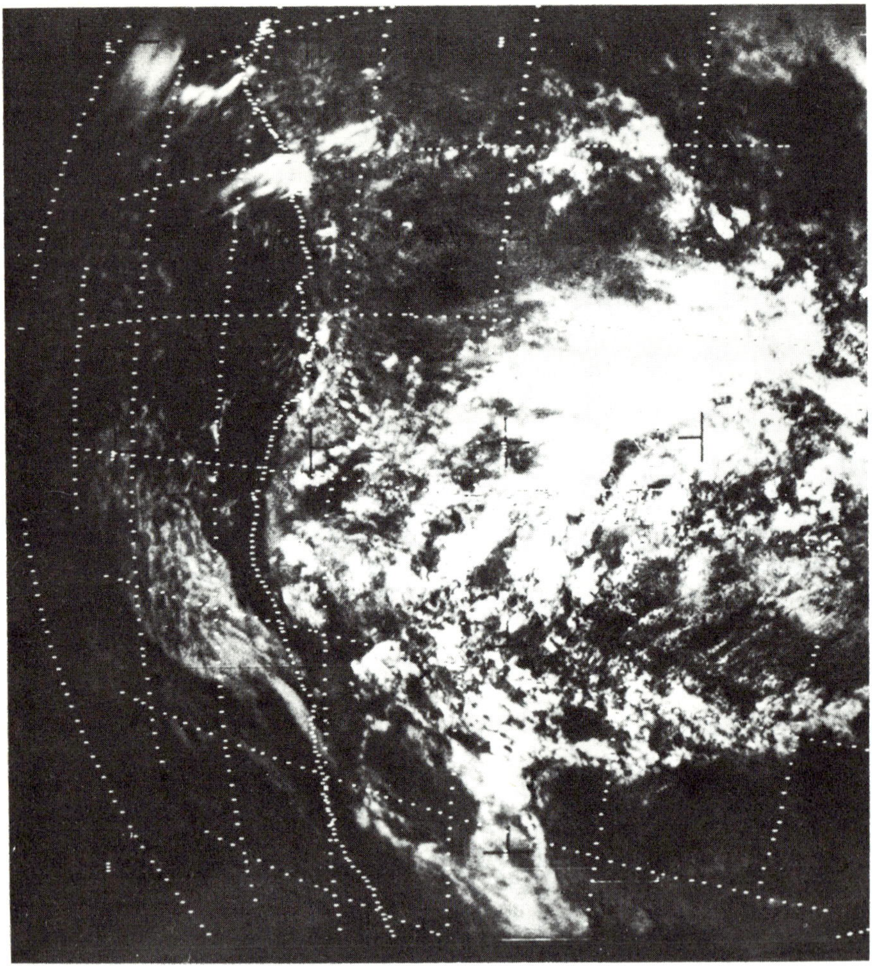

The Gulf Stream boundary during the night of 22 June 1966. Note the dark area of the Gulf Stream beginning at Cape Hatteras in the lower left corner of the picture at 35°N, 76°W, and continuing northeastward to 70°W, parallel to a cloud band which obscures much of the area to the east. Note the dark circular area centred along 40°N at 67°W.

in cloud-free conditions. Satellites with high resolution infra-red cameras have also been able to detect areas of sharp thermal contrast which indicate the boundaries of ocean currents and upwellings. These satellite photographs showed temperature differentials in the water by varying tones of grey (the warmer the temperature the darker the tone). Using such a tonal scale as this, it is possible for a computer to plot temperature maps on a geographic grid, which can be contoured to provide greater detail than was originally present in the photographs.

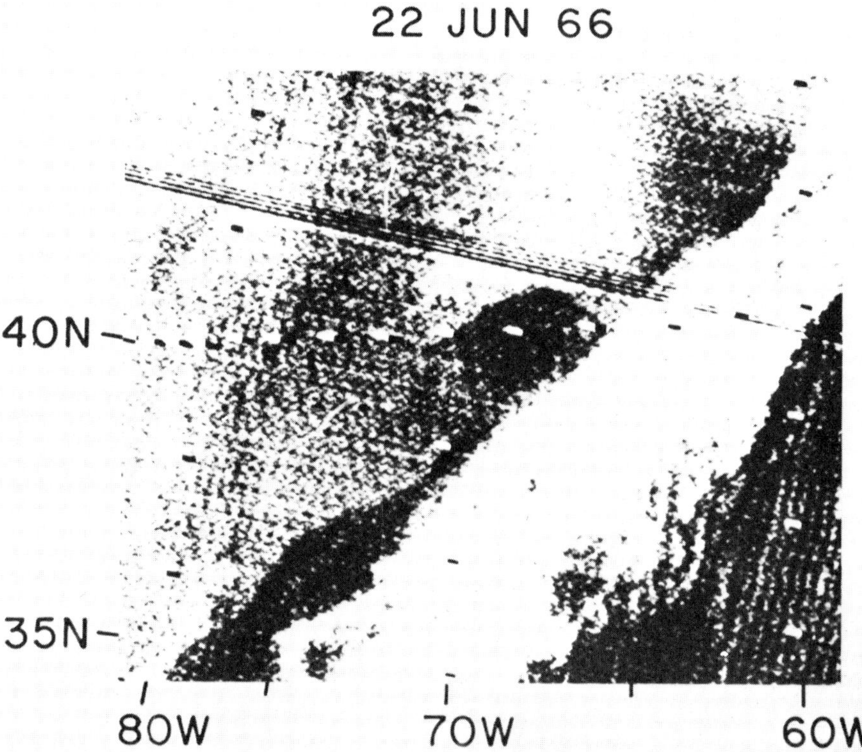

A section of the Gulf Stream boundary during the night of 22 June 1966.

The meteorological satellite Nimbus 2 was used to study large-scale fluctuations in the northern boundaries of the Gulf Stream Drift, just west of Cape Hatteras. The fluctuations observed were compared with those taken by a ship which had been measuring the 15° isotherm at 200 metres and that of an airborne radiation thermometer tracking temperature discontinuities between the continental shelf water and the warm current. Although there was a degree of uncertainty in the location of the boundary because of distortion in the satellite pictures, the comparisons made suggest that the possibility exists of establishing continuity in the gathered data.

During the 174 days of orbit of the Nimbus 2 satellite it was possible to observe the Gulf Stream, at least in part, nearly 50 times. It was also possible to observe other ocean currents and pronounced sea-surface temperature patterns with the infra-red cameras. Cloud cover retards optical recording of ocean-surface data, but it is possible to take advantage of the overall cloud patterns, and to deduce from them ocean temperature contrasts. The ATS 1 was able to photograph the planet from earth-synchronous orbit, and such photographs from high altitude permit the detection of variations in the cloud patterns related to the ocean surfaces beneath. By employing similar but perhaps more sophisticated methods,

The surface of the earth seen from a satellite.

it might be possible to perceive the earth's ocean as one comprehensible system.

Oceanographic forecasting up to now has had to depend on inferring sea-state from wind speed reports. Were it possible to observe the ocean in all weather conditions, accurate predictions for ships and meteorological stations could be provided.

All the methods employed to define sea-state from orbit are based on variations in the reflective properties of the ocean surface caused by changes in the wind speed and surface conditions. Aircraft photographic data have been used to determine the scattering of sunlight at the ocean's surface (sun-glitter) and assess its relation to wind fields. Azimuth angles, and the surface inclination needed to reflect light from any point on the ocean's surface, are determined by superimposed grids on the photographs. Television data from meteorological satellites has provided pictures of the non-reflective portions of the ocean's surface (non-glitter) which may also be used to determine sea-state.

Cloud cover is a limitation to any optical system of determining sea-state; methods involving passive microwave radiation and radar reflectance are being evolved. Both are able to provide accurate measurements even under storm and cloud conditions. Experiments have shown that reflected radar waves may be

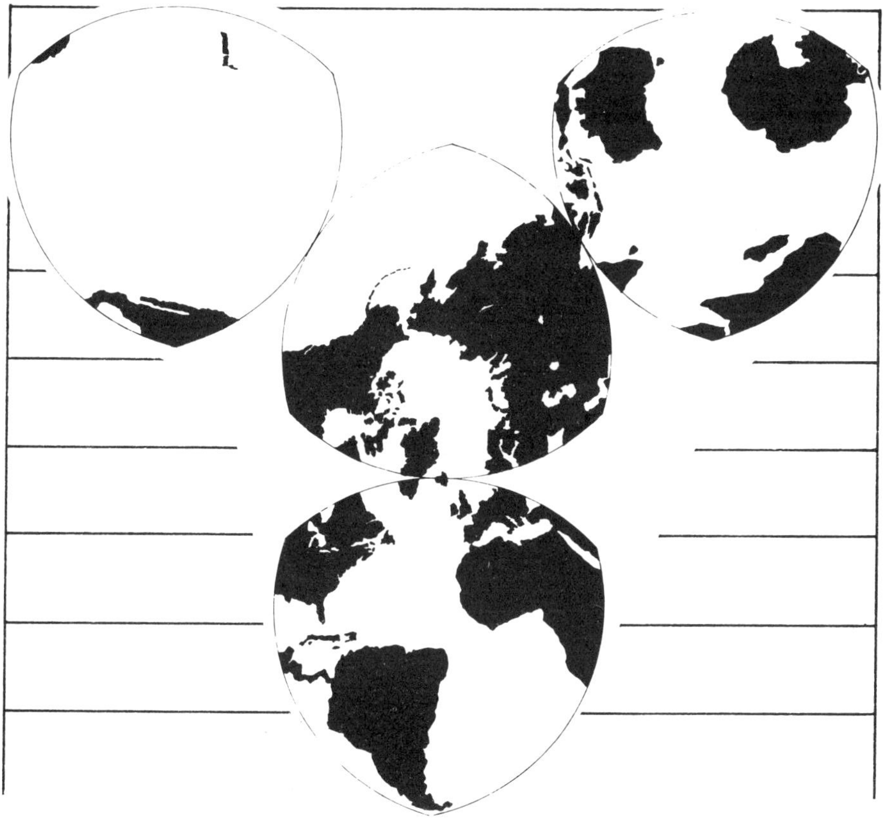

The projection of land mass and ocean over the planet's surface. The oceans tend to occupy the centres of the four triangular faces whilst land masses are generally distributed along the edges. A land mass on one side of the planet always has a very similar area of ocean on the opposite surface, for instance, the north pole which is ocean corresponds to the Antarctic land mass of the south pole. Similarly the whole planet can be divided into a predominantly land hemisphere (northern) and a predominantly ocean hemisphere (southern).

plotted against the angle of incidence at sea level to provide various readings for the degrees of turbulence in surface conditions. The relationship between surface roughness, temperature, angle of incidence, polarisation, and microwave radiation frequencies have been studied, and are equally unaffected by cloud cover.

On the Arctic and Antarctic oceans, ice coverage varies throughout the year. The distribution, behaviour, and variability of sea ice could be accurately observed from satellite data and photographs in all weather conditions.

The ability of satellites to give frequently a wide variety of data for the entire ocean could provide a unique method of assessing the abundance and distribution of potential food resources, if the fundamental relationship between environmental conditions and the migration of marine life can be accurately predicted. Temperature and ocean currents are already known to be closely

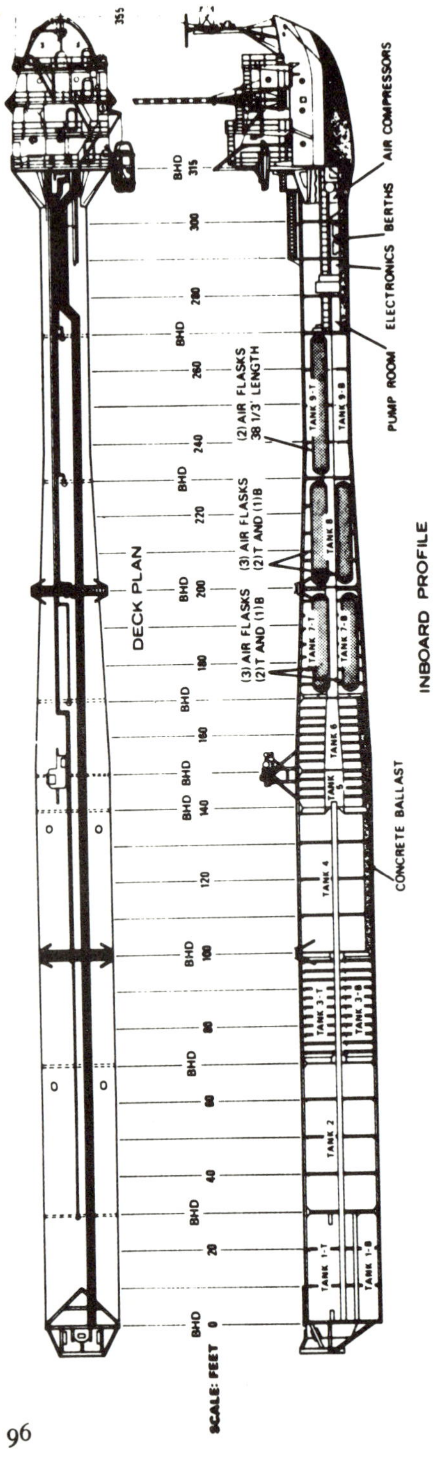

FLIP (Floating Instrument Platform) is the first of its kind developed by the Marine Physical Laboratory, Scripps Institution of Oceanography, to fulfill the need for an extremely stable and yet mobile platform, from which accurate acoustical measurements could be made at sea. With an overall length of 355 ft FLIP is towed in the horizontal position to the area where scientific operations are to be carried out. Upon arrival the tow line is cast off and the ballast tanks distributed throughout; after 255 ft of the vessel are flooded, the stern gradually sinks, causing the prow to rise. Ultimately the vessel is completely vertical with approximately 55 ft of prow pointing skywards, and the remaining 300 ft of the vessel underwater. The hull is narrowed down to a diameter of 12 ft 6 in for some distance above and below the vertical water line so that the rise and fall of the waves about the neck causes a very small percentage of change in the displacement. Consequently the vertical motions which occur are slight compared with those of a conventional ship. This stability, coupled with the unusually deep draft of FLIP in the vertical position, makes it an ideal station for a great number of oceanographic experiments. The floating platform displaces about 700 tons of water in the horizontal position, and about 2,200 tons in the vertical position. The hull consists of a number of sections; the lowest is a cylindrical section $12\frac{1}{2}$ ft in diameter. At the forward end of this structure is the 50 ft 'cobra head' shaped cabin. Four compartments in the cabin and four more in the adjacent cylindrical hull provide machinery space and working and living quarters for six crew members and as many as 15 scientists. There is no independent propulsion power other than

two small orientation motors, located below the vertical water line, which rotate the vessel about its vertical axis. These motors can be controlled from the gyro compass to maintain an accurate heading. Three diesel generators supply electrical power for all the ship's needs. Towing speed in the horizontal position is 10 knots, though it can be towed more slowly in the vertical position.

related to the movement of marine organisms. Schools of fish have been observed directly from the air, and under fine weather conditions could be observed from space. Another method of determining location, movement, and quantity of fish is based on the detection of various fish oils by spectroscopy. The fish oils absorb different wavelengths of light in the ultra-violet and infra-red range of the electromagnetic spectrum.

World-wide networks of surface stations and satellite systems that will provide the physical data by which the functions of the ocean as a whole unit may be observed in 'real-time', and the information relevant to the utilisation of the ocean's resources forecast, is being postulated in an Integrated Global Ocean Station System. This system is being developed by the Inter-Governmental Oceanographic Commission. A large network of satellite-interrogated buoys and advanced ocean data acquisition systems (ODAS) will be incorporated within it.

Man will have to utilise the natural, or artificially modified biological features of the ocean environment in order to meet the demands of a growing world population. The development of this immense area of the planet will not depend on the extension of existing methods of fishing or general exploitation, but on

A model of a tetrahedral floating city conceived by Buckminster Fuller.

Above left: A floating island complex conceived by French architect Albert and co-developed by Jean-Michel Cousteau.

Below left: A floating island complex conceived by Keith Critchlow as part of an elaborate ocean farming system.

radically new concepts such as undersea farms; the breeding of hybrid species of fish; artificial reefs and floating island complexes. Progress in these areas requires more sophisticated knowledge of predictive ecology than exists at the moment. The first experiments in this area will suffer the difficulties of any pioneering efforts, but the material rewards will be great.

The most dangerous threat to the successful use of the ocean as a food resource is the increasing pollution of the open ocean. Obviously, waste products must be disposed of and the ocean is quite capable of handling much of these. However, the increased dumping of contaminants, such as the chemicals of biological warfare, and radio-active isotopes and waste, in the firm belief that the deeper they are dropped the faster they will disappear, will eventually prove catastrophic. This 'out of sight out of mind' attitude will present future generations with the grotesque problem of a completely polluted and virtually useless ocean.

Over the ages in which sunlight has triggered the processes of life, living organisms have modified the terrestrial environment by selecting particular

wavelengths of sunlight which are most compatible with those processes. Before the origin of life, much more of the sun's radiation reached the surface of the earth than it does now. This is probably because the atmosphere at that time contained very little oxygen, negligible amounts of ozone, and very little carbon dioxide. A great deal of the sun's infra-red and ultra-violet radiation must have reached the surface of the earth during that period.

Some of the short-wave radiation, operating in the lower parts of the atmosphere, and probably in the surface layers of the oceans as well, must have been important in activating the synthesis and interactions of organic molecules which formed the prelude to the eventual emergence of the first living organisms. These organisms, coming into an anaerobic environment, and surrounded by the organic matter that had accumulated over the previous ages, most likely lived by fermentation, and this process must have produced as a by-product very large quantities of carbon dioxide. Part of this remained dissolved in the ocean and part entered the atmosphere.

Eventually the availability of large quantities of carbon dioxide, much greater than those present in the atmosphere today, made the process of photosynthesis possible. Any attempt to understand the basis of life on earth must inevitably come back to photosynthesis. This process enables plants to grow by utilising carbon dioxide (CO_2), water (H_2O), and small amounts of various minerals. Photosynthesis is the one large-scale process that converts simple, stable inorganic compounds into the energy-rich combination of organic matter and produces oxygen as a by-product, and consequently makes life on earth possible. Photosynthesis is the energy source of all living matter on the earth, and of all biological energy. The process can be simply illustrated by the formula:

$$CO_2 + H_2O + \text{sunlight} \rightarrow (CH_2O) + O_2 + 112{,}000 \text{ calories of energy per gram molecule.}$$

Photosynthesis began to remove carbon dioxide from the atmosphere, fixing it in organic form. At the same time it began to produce oxygen, which thus became established in our atmosphere. As the oxygen accumulated, the layer of ozone that formed in the upper atmosphere – also by a photo-chemical process – prevented further short-wave radiation from the sun from reaching the surface of the earth. The exclusion of this antibiotic radiation allowed living organisms from the oceans to move out on to the land.

The presence of oxygen also led to the development of cellular respiration, which involves gas exchanges that are just the reverse of those of photosynthesis. Eventually, photosynthesis and respiration came into approximate balance, as they have since remained.

Life, as a coherent natural event on a planet like the earth was inevitable. Similarly the whole of organic evolution lies within the overall order of nature. Astronomers have reason to believe that a planet of the size of our earth, with its temperature, and lighted by sunlight, is a reasonably rare event in the universe. Despite this, the universe is so vast that a conservative estimate would indicate that there are at least 100,000 planets like earth in our galaxy alone.

Experimental artificial reef designed by the Musée Oceanographique de Monaco, and section of the artificial reef (below).

About 100 million galaxies lie within range of the most powerful telescopes, so that throughout observable space there are quite probably 10 billion planets like earth. To place the origin of life within the concept of natural phenomena is to imply that life as we understand it probably exists on some, or all, of this vast number of similar planets.

Certain geological phenomena give simple organic compounds, for instance, volcanic eruptions bring metal carbides to the surface of the earth where they react with water vapour to give the simple compounds of carbon and hydrogen. A simple experiment of circulating a mixture of water vapour, methane, ammonia, and hydrogen (all gases believed to have been present in the early atmosphere of the earth), continuously for a week over an electric spark was performed by S. L. Miller in 1955. The circulation was maintained by boiling water in one part of the apparatus and condensing it in another. At the end of a week the water was analysed by the delicate method of paper chromatography, and it was found to have acquired a mixture of amino acids. Glycine and aline, the simplest of amino acids most prevalent in proteins, were identified in the solution. Assuming that all these factors and others which have yet to be discovered, together eventually give the process of respiration which is the energy exchange method of all animal life, then:

$$C_6H_{12}O_6 + 6O_2 \rightarrow 6CO_2 + 6H_2O + \text{energy}.$$

The burning of 180 grams of sugar during cellular respiration yields about 700,000 calories, as compared with approximately 20,000 produced by fermentation of the same quantity of sugar. This process of combustion extracts all the energy that can possibly be derived from the molecules which it consumes. With this process at its disposal, the cell can meet its energy requirements with a minimum of expenditure of substance. It is a further advantage that the products of respiration, water and carbon dioxide, are innocuous and are easily disposed of in any environment. Æons ago, this overall polymerisation sequence went on selectively and efficiently in the earth's atmosphere and led to the formation of a protein-dominated layer that was concentrated in the oceans of a then sterile planet. It is quite likely that such processes are at present occurring on Jupiter which possesses an atmosphere of methane and ammonia, and perhaps will eventually form oceans there, which are the precursors of the origin of life.

It is quite conceivable that there are not many ways of composing or constructing living organisms that are fundamentally different from those already known to man. Within conditions where life is possible, it must eventually arise, and develop into forms which may differ in detail from those we know, but could include many which look quite familiar. Fortunately man is not alone in the universe and does not bear all the enormous responsibility for the continuance of life, the most sophisticated and complex state of organisation that matter has achieved.

The wreck of the nuclear submarine USS Thresher *photographed from the bathyscaphe* Trieste; *a similar accident to a ballistic missile submarine could cause serious radioactive pollution of the oceans.*

Below: U.S. Navy crew members form the equation '$E = mc^2$'.

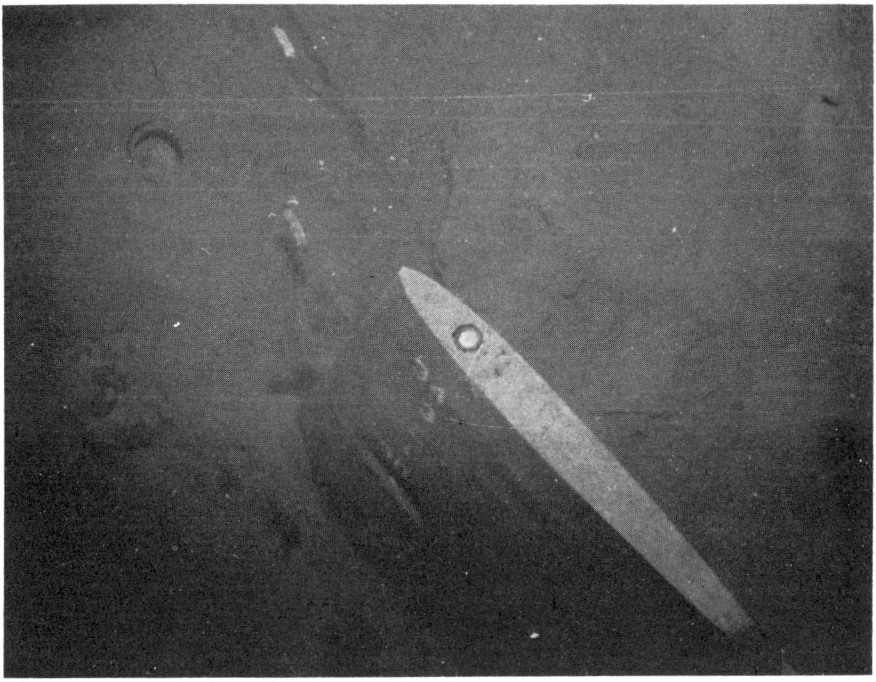

E=mc 2

Route of U.S.S. Nautilus *on a submerged crossing of the North Pole.*

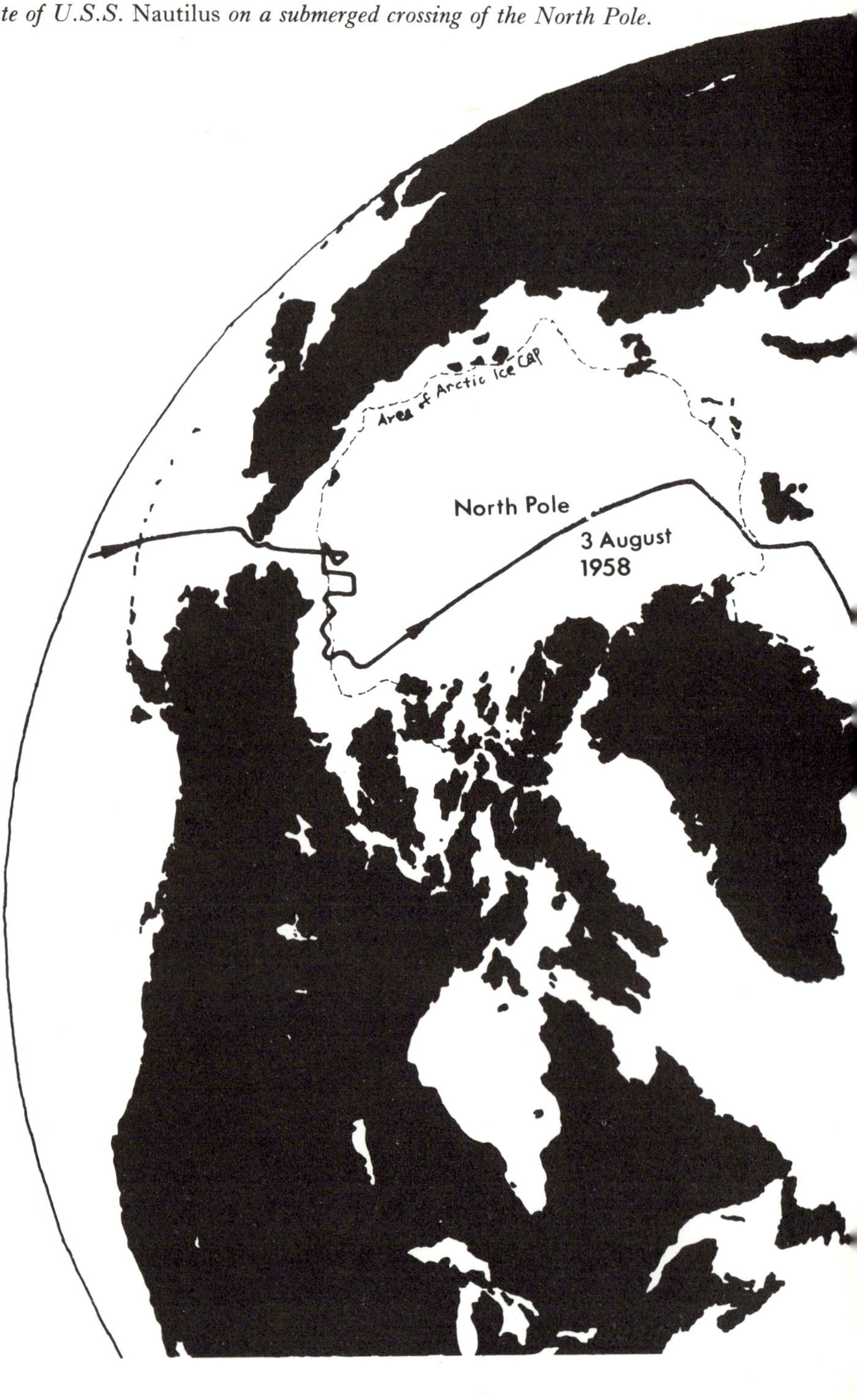

f
utilus
TLANTIC
OCEAN

September 1990
Natural History
The Way It Was:
Clean Air,
Clean Water
Will you
read our
September, 1990 issue
with a sense of nostalgia?

Appendices

Above left: Polaris missile submarine H.M.S. Resolution.

Below left: 1990: 'The Way it Was: Clean Air, Clean Water'.

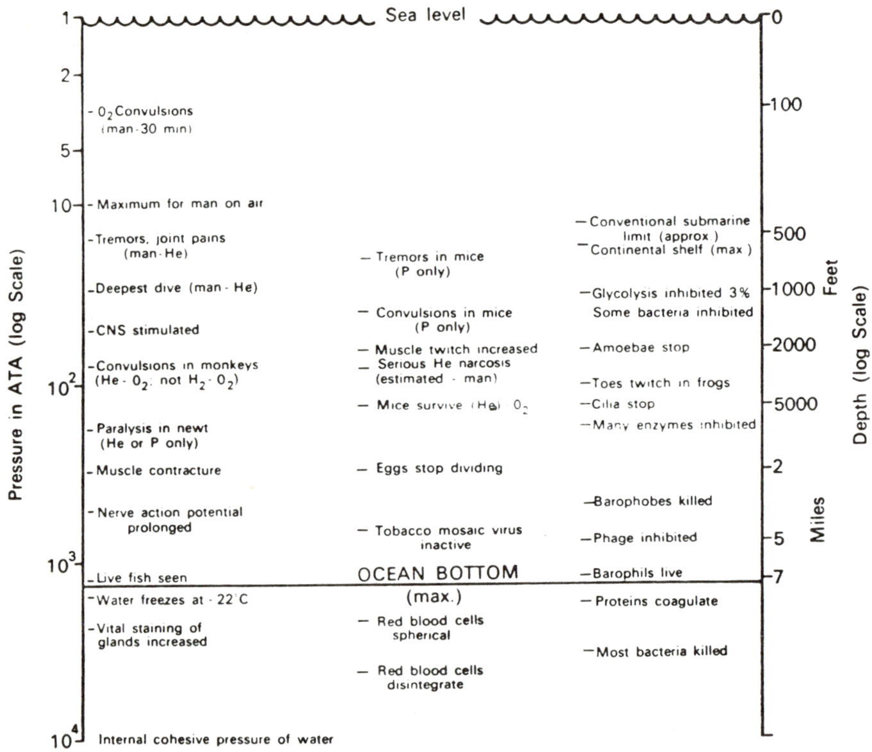

The effects of hydrostatic pressure.

Appendix 1

Man's adaptation to undersea life. Notes from a paper presented at the C.R. Académie des Sciènces, Paris, 25 April 1966, by Jacques Chouteau, Jacques-Yves Cousteau and Jean Alinat.

A series of experiments have been undertaken on prolonged dives, from 10 to 120 metres deep, in compressed air or a mixture of oxygen/helium, and based on experiments with animals to a depth of 200 metres. They have enabled us to define and calculate the 'coefficient of the useful duration of a dive'.

The problems set by life in a pressurised atmosphere – and thus by an underwater dive requiring the use of a compressed air mixture – are numerous. Some of these problems have come to light, been investigated and, in some cases, solved, following extensive research carried out by Paul Bert and numerous tests that followed. The limits in depth – similarly in pressure and duration – are essentially dictated by narcotic properties formed by a gas mixture breathed under pressure, and by phenomena accompanying decompression.

A depth limit of about 40 metres is imposed by the first of these factors even though at an absolute pressure of 5 kg cm^{-2}. Replacement of nitrogen by a lighter gas, such as helium or hydrogen, greatly extends the limits of this narcosis, although how much is not specified.

The second limitation arises from the fact that the greater the depth (or the pressure) the less the duration of the immersion permitted without special precautions. In order to increase the duration of immersion at a given depth, it is necessary to pass through the decompression chamber on return, for a length of time that increases with the amount of the saturation of the organism in gas dissolved at the same pressure of the stay at the bottom.

The following gives an idea of a relationship:

$$\frac{\text{length of immersion (thus of work)}}{\text{length of immersion} + \text{duration of journey to surface}}$$

which represents the *coefficient of useful duration of a dive.*

Clearly, this relationship will diminish near to saturation point (since the time taken for surfacing increases), then it will approach unity once saturation point is reached, the length of decompression staying constant. Decompression that limits short-stay immersions thus becomes less and less important, to the point of becoming negligible for a lengthy immersion.

Although many experiments have been devoted to short but deep dives, the study of prolonged dives (*saturation dives*) is only very recent. Two preliminary experiments were carried out on animals and showed the limiting and even lethal effect of hyperoxia on the one hand, and the increase in the density of the respiratory mixture on the other.

The purpose of these experiments known as *Precontinent Experiments* (EP) was essentially:

1) To calculate and make a practical study of the 'coefficient of the useful duration of a dive'.

2) To lay down the conditions for prolonged dives, in particular the nature of the respiratory mixtures to be used, and their drawbacks.
3) To show the absence of harmful effects, long and short term, of these saturation dives in the conditions described here.

EP I or *Diogenes* with 2 men (14–21 September 1962): Marseilles area (open circuit ventilation). Decompression: 3 hr; O_2 for respiration at bottom.

EP II, with 6 men and 2 men respectively: 9.5 m (15 June to 14 July 1963): Red Sea, Shaab Rumi (same as EP I); at 25 m (5–13 June 1963): same environment (closed circuit). Decompression 3 hr 80 per cent O_2 base for rebreathing chamber from 20–15mn and 30–12mn, one night in the main house at 9.5 m.

EP III, with 6 men each time: at 25 m (28 August–2 September 1965): Monaco Harbour, at 100 m; 17 September–17 October 1965: Cap Ferrat.

All the other experiments (with sheep, goats or men respectively) from January–August 1965: (high pressure) chamber from OFRS Marseilles. (In these experiments, as in EP III, decompression follows the curve of desaturation of the tissues over the longest period: 120 mm with passage O_2—He to the air about 4 atm.)

The composition and specific volume of the atmosphere breathed is indicated here, and the partial compression of O_2, N_2, CO_2 and He.

Experiments EP I, II and III took place under water, the subjects living in habitats (underwater houses) exposed to the open sea, and at the same pressure as the surroundings. The subjects, fitted with breathing apparatus, carried out various tasks for three to five hours a day (photographing, filming, collecting samples, looking after an oil-well) and at depths greater than that of the habitat (at an absolute pressure of 30–50 per cent greater, permitting re-entry without decompression into the house, or at an even greater pressure but for a limited time, or with re-entry via the decompression chamber).

The experiments carried out in an underwater chamber on both animals and men, in addition to a physiological aim, have enabled various tests to be carried out on the degree of oversaturation permitted for decompression.

In conclusion, all the tests undertaken have shown conclusively:

(*a*) That life is possible under pressurisation, on condition that the partial compression of the oxygen in the respiratory mixture used resembles as closely as possible that of the air breathed under atmospheric pressure (210 g cm^{-2}) on the one hand; and, on the other hand, that the density of the respiratory mixture is lower than that of the compressed air breathed under 2–3 kg cm^{-2} ($2{\cdot}6$ to $3{\cdot}88 \,.\, 10^{-3}$ g cm^{-3}). This is produced by substituting helium for nitrogen.

(*b*) That it is possible to carry out effective work diving from these underwater habitats.

(*c*) The complete absence of distress experienced by the animals and men; or, more precisely, the absence of pathology attributable to life under pressurisation during, or after, the experiment.

(*d*) The satisfactory nature of the forms of decompression used. During the tests in the underwater chamber on animals and men, more rapid decompression was tried which allowed assessment of the margin of security provided by the methods of decompression used for long immersions for animals and men.

Appendix 2

Precontinent 1
The EP I experiment, once known as *Diogenes*, took place between 14 and 21 September 1962, at Marseilles. Two men lived for seven days in a submerged habitat, 25 m^3, anchored to the sea bed. The aquanauts could get in and out through an opening in the lower generator. This opening also allowed pressure to be balanced at 2·06 Ats with partial oxygen pressure (P10) of 316 Torr.

Compressed air in open circuit coming from 400–500 l p min. The men worked under the water for five hours at a time at depths of 5–25 m, temperatures ranging from 21°–16°C, food coming in watertight containers from the surface.

During their stay they showed slight adaptation syndrome, due to anæmia arising from the light hyperoxia that they were undergoing. This last point confirmed previous animal experiments and explained $P10_2$ less than 309 Torr of earlier experiments. At the end of their stay, although no decompression was necessary, the men came up, having breathed a mixture of 80 per cent O_2 and 20 per cent N_2 intermittently for 3 hours and a symbolic rest at 3 m.

Precontinent 2
The aim of the EP II experiment, 14 June–14 July 1963, on the reef of Shaab Rumi in the Red Sea, 23 nautical miles north of Port Sudan, was:
1) To stay down longer than before (30 days), at a similar depth 9·50 m (1·95 Ats) using more divers (6 men) less carefully selected.
2) To set up a station at greater depth 26·5 m (3·64 Ats) with two men for 6 days, in an atmosphere 50 per cent air and 50 per cent helium, in closed circuit with regeneration, relying on the safety of a better dwelling unit.
3) To demonstrate the possibility of assembling underwater dwellings, far away from bases, using a surface support starting from submerged dwellings or hangars, as many items of apparatus as possible were used: diving saucers, scooters, television, telephone (ultrasonic), movie photographic equipment, inflatable equipment, etc.

The submarine village was made up of a large house for six men fed from the surface with compressed air; a maintenance boat-house for the diving saucer, similarly fed with pressurised air at 9·50 m; the small house for the two men in closed circuit at 26·50 m; and various other sheds or shelters. All this was installed from the ships *Calypso* and *Rosaldo*, which provided energy, compressed air, food and all other necessary services for the aquanauts.

The aquanauts dived from 3–5 hours a day on various missions (oceanography, and filming *Le monde sans soleil*) in a temperature of approximately 28°C; usually at 25 m, and for short excursions at 50 m, for those from the upper dwelling unit usually at 50 m, and occasionally at 80–100 m. All this was carried out without any need for decompression. No specific problem confronted the divers and there were no incidents as in the previous experiment. Only difficulties traceable to heat and humidity were observed by the aquanauts.

For the occupants of the upper dwelling unit, decompression was the same as that in *Precontinent 1*; the two aquanauts in the dwelling unit at 26·50 breathed a mixture of 50 per cent O_2 and 50 per cent N_2 for three hours, making a landing

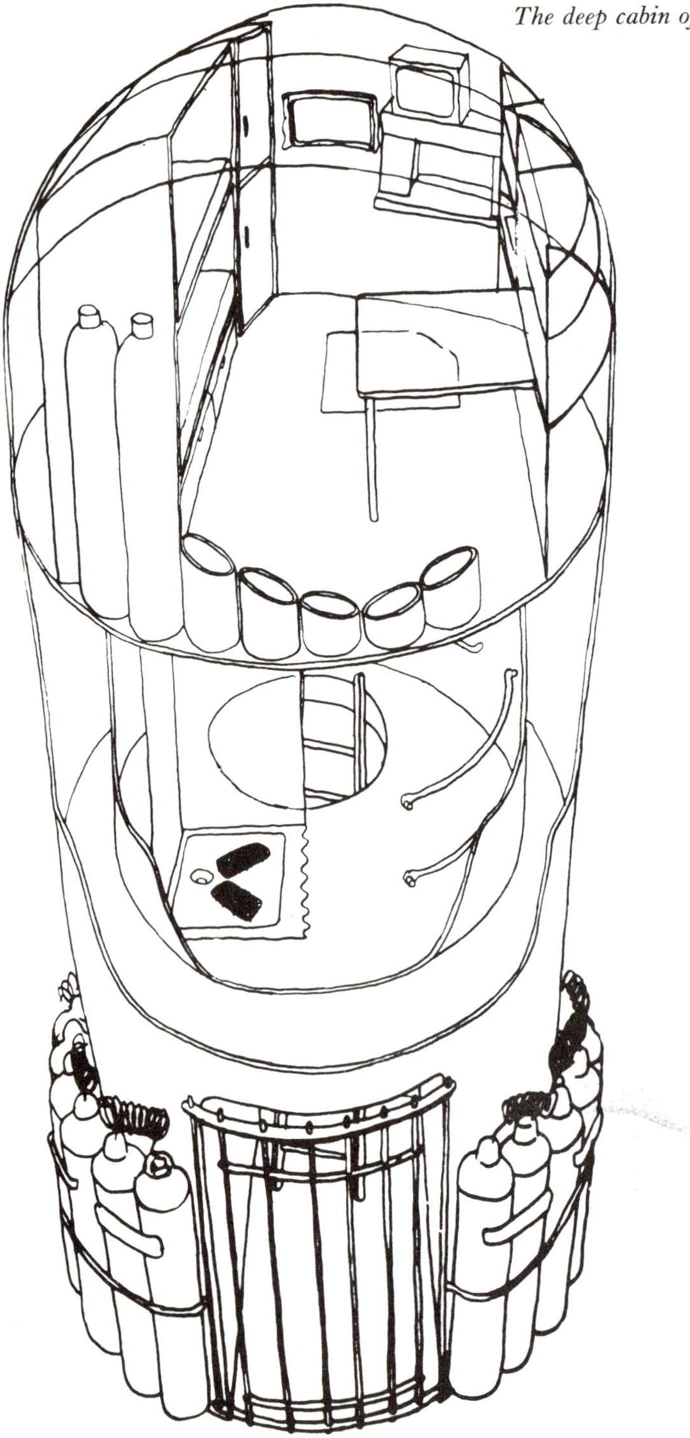

The deep cabin of Precontinent 2.

after 20 minutes, at 15 m in fresh air, a landing of 30 minutes at 12 m, and passing a night in the house at 9·50 m before coming up to the surface.

The diving saucer SP 350 carried one pilot and one observer. Its operational depth is 350 and its cruising speed powered by two hydrojets is 0·7 knots (outboard motor of 2 CV). The self-contained electrical system lasts for three hours, and the breathing system 24 hours. The weight is 4·5 tons in all, enabling it to be transported as air cargo.

Precontinent 3

EP III took place between 17 September and 17 October 1965, off Cap Ferrat at 100 m (depth 10·95 Åts) using a mixture of oxygen and helium (2–3 per cent O_2) with six men.

The point of the exercise was to try to increase the depth, using test animals (goats, sheep) and men with little psychological disturbance. The pressure exceeded 11 atm. After a trial period of 4 days at 25 m, six men lived at 100 m in the spherical house. The aquanauts were put into compression in the dwelling unit in the port of Monaco, then towed out and submerged.

The dwelling unit was self-contained with regards to feeding and atmospheric renewal (conduction of oxygen, compensation for helium losses, absorption of CO_2 and the remnants of lime, control by spectrography and chromatography, etc.); it was connected to the surface for the supply of electrical energy and an information transmitter. The diving saucer and lift, made up of watertight containers, ensured further contact. The aquanauts worked four hours a day at a depth of 130 m (oceanography, upkeep of an oil-well head, connected by a tube maintaining a mixture of air from the dwelling unit).

At the end of their stay, the dwelling unit was closed and resurfaced by the aquanauts, decompression starting during the tow back to Monaco harbour. Decompression took 84 hours.

Precontinent 2: *diver.*

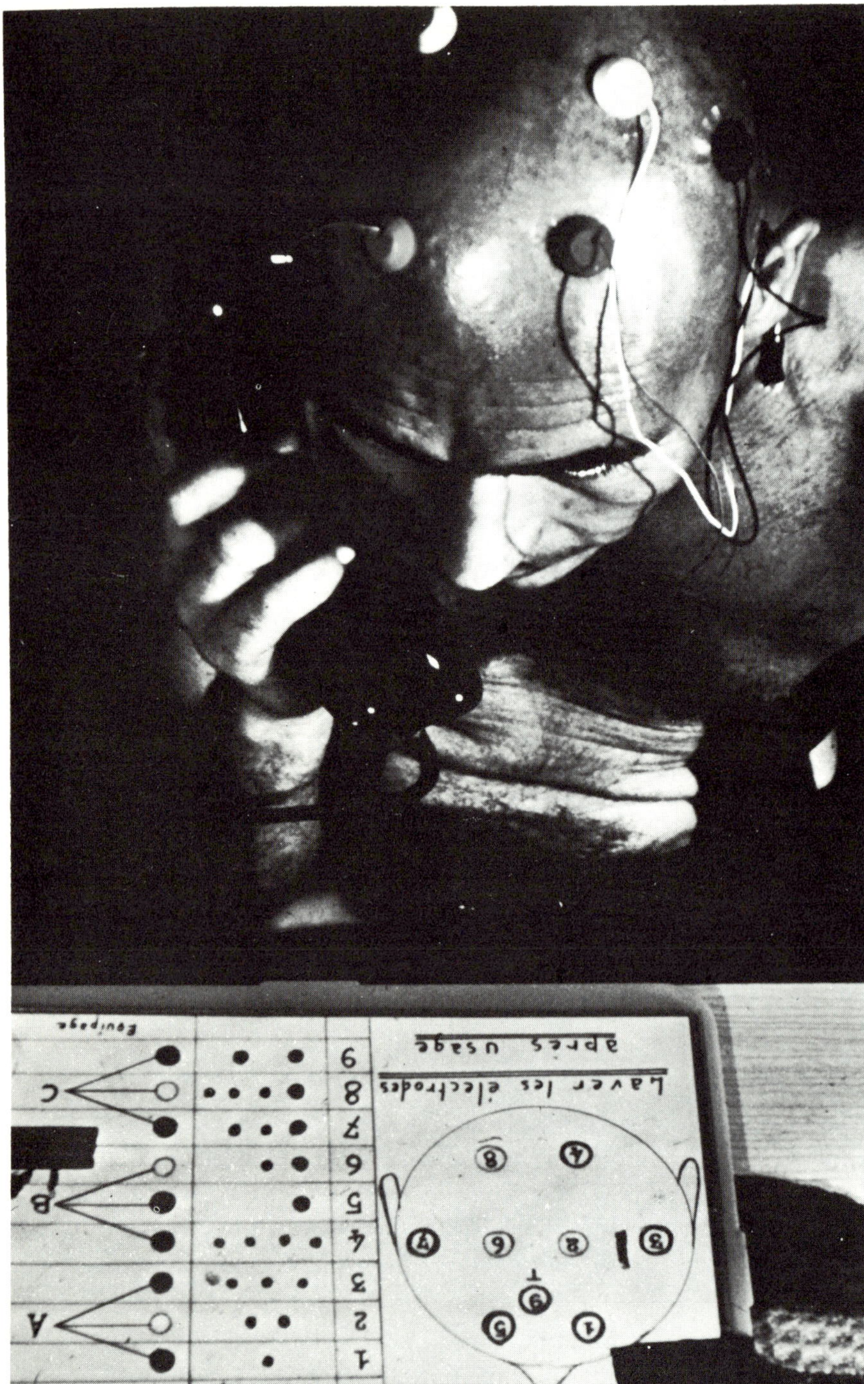
Laver les électrodes
après usage

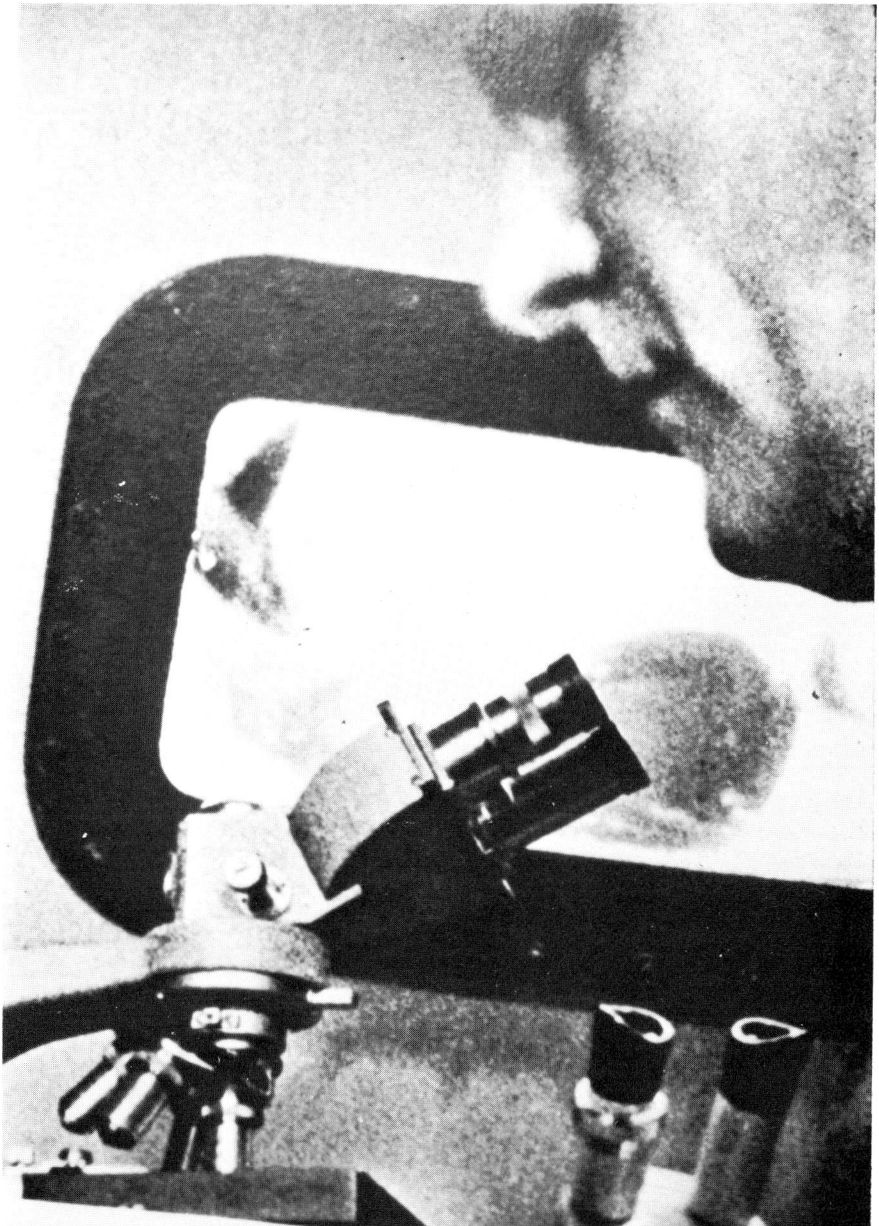

Precontinent 2: *biology laboratory.*

Left: Electrodes on the head of Precontinent 3 *oceanaut to measure any disturbances in the electrical activity of the brain.*

TRASH

Appendix 3

Aquanaut daily routines and physiological evaluation of Sealab 2, by Malcolm Scott Carpenter and George Bond of the United States Navy. (Notes from Office of Naval Research Report ACR 124.)

Once the lab was reasonably habitable for 10 men, all of the spare time in the water was devoted to the scientific programmes and equipment evaluations. These activities included the erection of the strength-test platform and associated torque wrenches, the two-hand co-ordinator, the current meter, underwater weather station and sound range, visual acuity range, stationary target array, water clarity meter, pneumofathometer, fish cages, homing beacons, compass rose, external TV cameras, bioluminescence meter, foam and salvage project equipment, bottom current trailers, underwater studgun equipment, photo and diving lights, bathythermograph, wave gauge, and antitorque underwater tool test equipment.

Before and after each dive, strength and manual dexterity tests were performed in the water with the aid of equipment designed specifically for this purpose. Straightening the outside gas mixture hoses to supply the divers had to be done almost daily. Re-supply through the system of pots and baskets also consumed too much of the divers' time.

During the third team's tenure, the storm of activity centred around re-supply abated somewhat because of the installation of a high-pressure helium-mix line in the lab. This line, supplied by pumps on the surface ship, permitted re-charging of the Mk VI bottles in the lab instead of sending them topside for refilling. It not only reduced the workload on the men, but was also kinder to the equipment.

A watch schedule had been set up prior to the first dive, and it went into effect immediately after the lab was occupied. The schedule provided two men on watch at all times during the working hours, and one-man watch sections during the sleeping hours. The responsibilities of the working-hours watch section included keeping the Log, handling communications with topside, preparing and cleaning up after the meals, and staging the divers. It was more a rule than an exception that these tasks had to be performed concurrently; and frequently the duty section found it necessary to stage divers, keep the Log, cook, communicate, and clean up, all at the same time.

The sleeping-hours watch section, although only one man, was not nearly so hectic. There was time to engage in a little idle chatter with the surface, make the daily report activities, finish off the Log, and set up four or six Mk VI breathing rings for the next day's sorties.

The daily schedules changed each day, as did the 'Plans-of-the-Day' (P.O.D.), until a smooth operation evolved out of the first trial-and-error week of activity on the bottom.

Physiological Studies

It was considered prudent to pursue a course of selective monitoring of personnel

Left: Sealab 2 *being lowered onto the deck decompression capsule on staging vessel.*

of each of the three teams for overall safety of the operation. Previous meticulous physiological studies, conducted through the *Genesis* series, and subsequently in *Sealab* 1, had indicated that the great majority of physiological parameters examined would show no significant change under conditions of high pressure and exotic gas mixtures. It was therefore considered adequate to monitor only those vital functions of the human subjects which might assist topside control in medical management of the experiment.

A programme was established to provide for intensive physiological sampling from two or three aquanauts of the first two teams on a daily, alternate daily, then each-third-day basis. Sampling included extraction of about 20 cm of blood daily for topside analysis, with appropriate quantities of urine and saliva for additional inspection. In addition, the majority of aquanaut personnel participated in daily studies of pulmonary function, electro-cardiographic recordings, body-temperature control, exercise tolerance, and a number of routine physical tests.

By and large, the results of tests on selected subjects in the three teams of *Sealab 2* were essentially negative.

Author's Note

On 15 February the *Sealab 3* habitat was lowered 610 ft to the ocean floor off San Clemente Island, California, to begin the third saturation dive experiment of the U.S. Navy. The experiment had suffered unfortunate delays and was two years behind its schedule deadline.

The *Sealab 3* submerged pressurised and unmanned at a planned rate of 4 ft per minute. Having arrived at the bottom, the surface, monitoring on board the *Sealab* support ship, *Elk River*, indicated a helium leak and a consequent imbalance of the internal atmosphere. On 17 February four divers were sent down in the personnel transfer capsule to investigate and repair the leak in the habitat. Aquanaut Berry Cannon and his co-diver were finning towards the habitat when TV monitors and co-divers observed Cannon swimming awkwardly and in obvious difficulty. Despite being immediately returned to the PTC and receiving heart massage and resuscitation, Berry Cannon died of heart failure by the time the capsule reached the surface.

The autopsy revealed that Cannon had died of cardio-respiratory failure due to carbon dioxide poisoning. Unfortunately, how the lethal quantity of CO_2 was breathed was not determinable, and the experiment was terminated.

Below: Sealab 2 *begins descent.*
Right: Experimental plant study in Sealab 2.

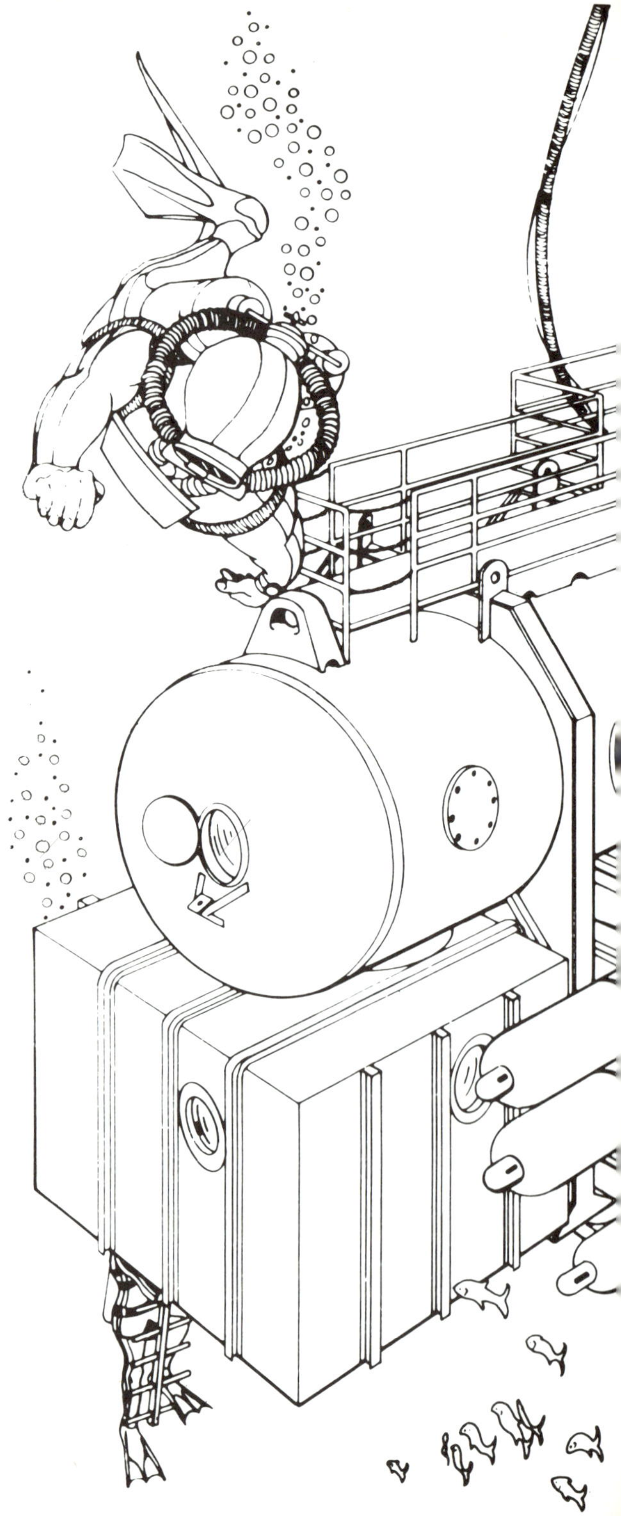

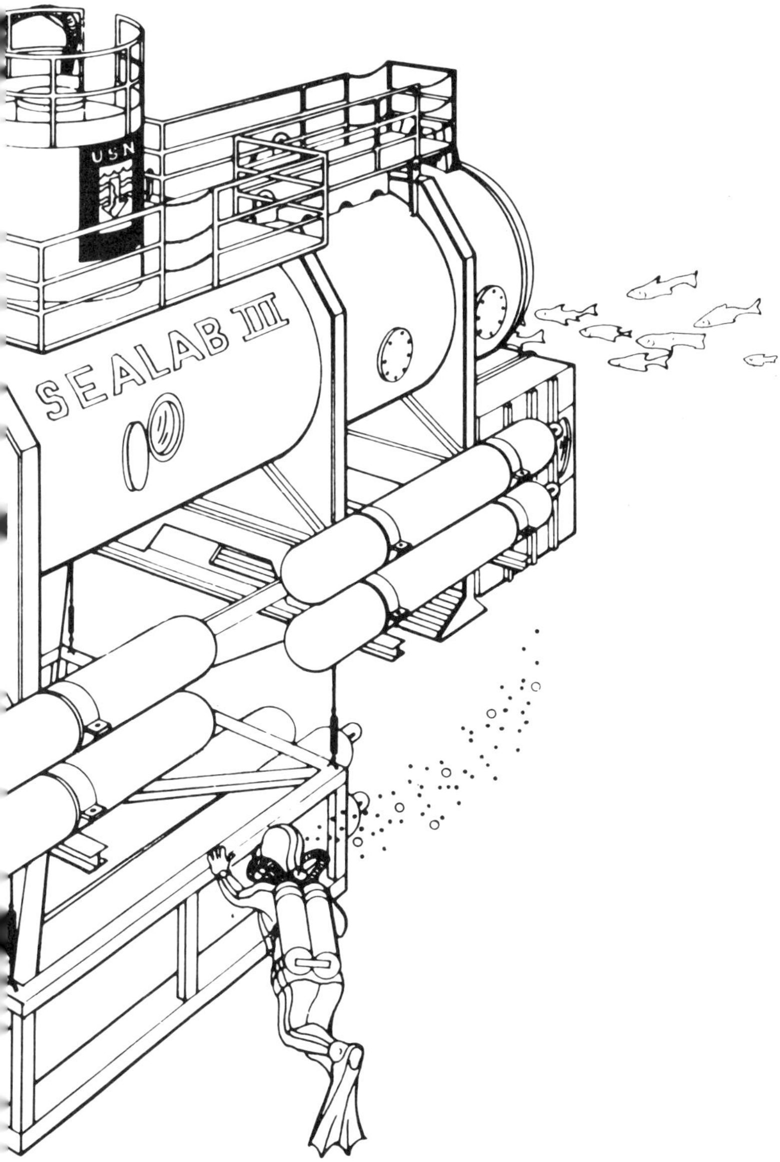
USN
SEALAB III

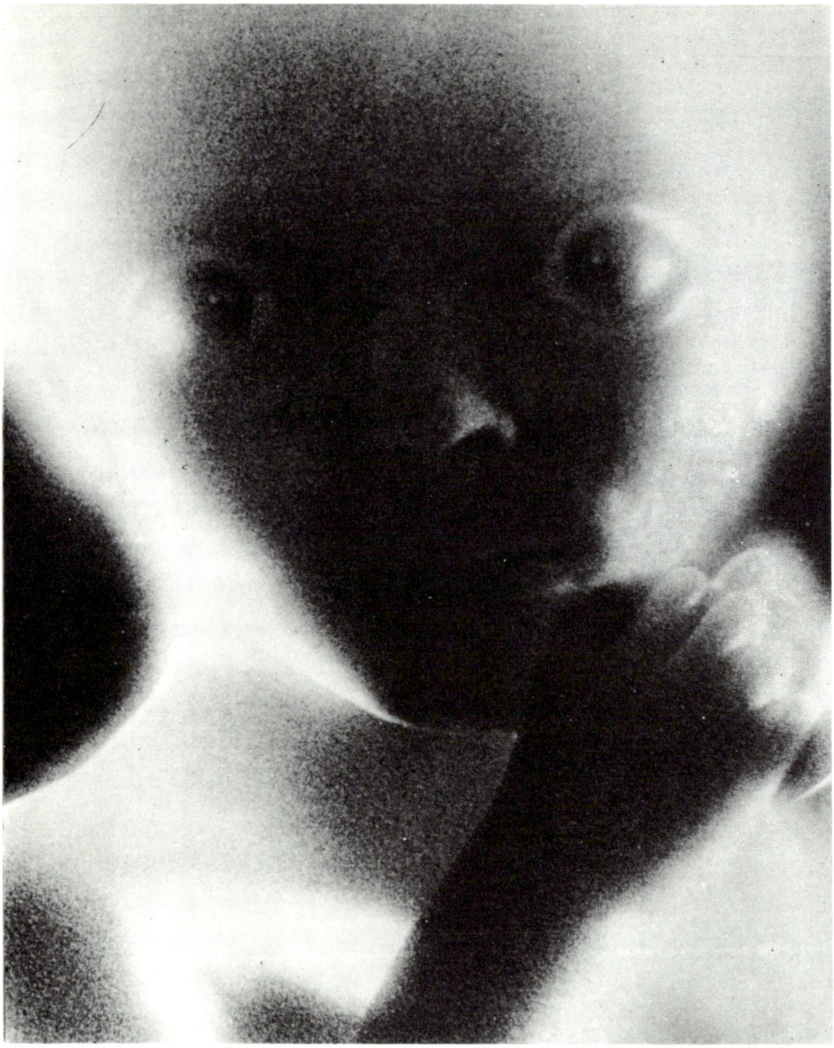

Starchild (*from* 2001 – A Space Odyssey)

Picture Credits

Much of the illustrative material has come from the following people and organisations, without whose help this book would not have been possible. J. D. Bernal, Liquid Group, Birkbeck College: British Museum: Jean-Michel Cousteau: Keith Critchlow: J. R. Geigy S.A.: G.E. Research and Development Centre, Schenectady, New York: Hulton Picture Library: Stanley Kubrick: Lockheed Missiles in Space: Stanley Miles, Staples Press: Ministry of Defence, Crown Copyright Reserved: N.A.S.A., Jacques Picard: Les Requins Associés: Reynolds Aluminium Corporation: Scripps Institute of Oceanography, University of California, San Diego: Paul Shepherd: Jacques Stevens: Tate Gallery, London: United Features Syndicate Inc.: U.S. Navy Deep Submergence Systems Project Office, Chevy Chase, Maryland: U.S.I.S.: Westinghouse Electric Corporation: The geographical projections are by the author.

The author wishes to thank Harper and Row for permission to reproduce 'Menfish' from Living Sea *by Jacques-Yves Cousteau.*

Bibliography

SVERDRUP, FLEMING and JOHNSTON, *Oceans, their Physics, Chemistry and General Biology*, New York, Prentice Hall, 1942.

Jacques-Yves COUSTEAU, with Frederic Dumas, *Silent World*, London, Hamish Hamilton, 1953.

Jacques-Yves COUSTEAU, *Living Sea*, London, Hamish Hamilton, 1963.

Jacques-Yves COUSTEAU, *World without Sun*, London, Heinemann, 1965.

N. R. DIBELIUS, A. DOUNOUCOUS and W. L. ROBB, 'Permselective membrane systems for underwater life support', G.E. Report, August 1965, Documenta Geigy Nautilus 1–7 J.R., Basle, Switzerland, Geigy S.A.

John L. MERO, *Mineral Resources of the Sea*, Amsterdam, Elsevier, 1965.

Stanley MILES, *Underwater Medicine*, London, Staples Press, 1966.

G. NEUMANN and W. J. PIERSON, *Principles of Physical Oceanography*, New York, Prentice Hall, 1966.

Robert STENUIT, *The Deepest Days*, London, Hodder and Stoughton, 1967.

Project Sealab Report ONR ACR 124, edited by Pauli and Clapper, Washington D.C., Office of Naval Research, United States Navy, March 1965.

Proceedings of the 3rd Symposium on Underwater Physiology, edited by Charles Lambertson, Baltimore, Williams and Wilkins, 1967.

Raymond VAISSIERE, *L'Homme et le Monde Sous Marin*, Paris, Larousse, 1969.

'Inner Space', special issue of *Architectural Design*, April 1969.

'Oceans', special issue of *Scientific American*, July 1969.

J. A. KYLSTRA, 'Land Animals in Hydroecology', in *New York State Journal of Medicine*, October 1969.

R. J. TAYLOR, *Water*, Unilever Educational Booklet, Advanced Series No. 5, 1966, revised edition 1969.

Index

Figures in italic refer to the captions to the illustrations